Warrior Dogs

Warrior Dogs

Getting to the Battlefield

LTC (ret) Richard A. Vargus

Red Engine Press
Fort Smith, Arkansas

Copyright © 2023 Richard Vargus

ALL RIGHTS RESERVED. No part of this book may be reproduced or transmitted in any form or by any means, electronic or mechanical, including photocopying, recording, or by any information storage and retrieval system (except by a reviewer or commentator who may quote brief passages in a printed or on-line review) without permission of the author.

Library of Congress Control Number: 2023945142

ISBN: 978-0-9885891-1-7

Edited by Pat Avery, Lura Patrick, & Joyce Faulkner

Cover art by Joyce Faulkner

Views Expressed Disclaimer

Subjects, thoughts, opinions, and even presentation of facts within this work reflect only the author's views and not the wider views of the publisher.

Dedication

September 11th, 2001 (9/11) will live forever in the hearts and minds of Americans. Our nation was attacked, maliciously and without provocation. First responders and innocent people across the spectrum of nationality, color, race, and ethnicity—simply going to work in the Greatest City in the world—were killed or injured. On that day, the life of America changed forever.

All New Yorkers thoughts were affixed on the rescue effort, hoping, and praying there would be survivors. Like the rest of the country, I was glued to the television, shocked and angry. I lived on Long Island, the hub that fed the city with workers traveling daily on the Long Island Expressway and Long Island Railroad. Thousands of people made the daily trek into the city from the "suburbs."

We took for granted that Long Island was the final approach for flights returning from Europe to JFK. The sound of large planes passing overhead was as normal as breathing. But after two of them struck the World Trade Center and all aircraft were grounded, the absence of aviation traffic overhead was deafening. Long Island remains the home of the largest community of NYPD, and FDNY first responders. The fate of fathers, mothers, grandparents, sisters, brothers, daughters, sons, and friends were in God's hands—and his infinite mercy.

One of those Long Islanders, a friend since high school was Firefighter Thomas Gambino, a nineteen-year veteran of the Fire Department City of New York (FDNY)—a member of the elite FDNY Rescue 3, one of the first units to enter the South Tower. Tommy—like so many of the brave men and women who ran towards the burning buildings—was just doing his job. No one that day professed to be a hero. That was because everyone was a hero! First responders, civilians, men, women, kids on the street, it didn't matter. New Yorkers showed the world the best of humanity. New Yorkers left "NO ONE BEHIND."

Tommy and his fellow FDNY brothers and sisters did what they were trained to do since the inception of the FDNY in 1865. New York's Bravest, knew the danger and the possibility that this would be their last call, but no one hesitated. They had a job to do. Not one of New York's first responders turned back or worried about their own safety. They marched toward the belly of the beast.

No one expected what came next. This was the World Trade Center. These buildings were indestructible. Then in horrific succession, the South and North Towers collapsed. The façade of the city changed forever. And amid the mass confusion—in a flash, the FDNY lost 343 souls. The eerie sound of their emergency beepers blared throughout the disaster area, which was, for many, the only part of their bodies ever recovered.

For days I meticulously followed the news, watching Long Island's WGLI Channel 12, one of the few stations still operational. The major networks moved all telecasts, since their transmitters, located on the top of the Trade centers, were destroyed. Every day I maintained a vigil, looking in Newsday, hoping and praying that I wouldn't see Tommy's name. Then one day, it appeared …Tommy Gambino Jr., FDNY Rescue 3. I was devastated.

Tommy was a dedicated husband and father…and an extremely devout Catholic. He loved being "on the job." For as long as I knew him, I never heard him say a bad word about anyone. Sporting that big broad smile, he was always happy to see you. A dear friend, he was the kind of person I aspired to be.

Tommy had much to live for, but he unhesitatingly ran into that building, never to be seen again. If Tommy had lived and had to make that life and death decision to run into the Trade Center, his decision would have been the same. All first responders and our men and women serving in the armed forces pledge themselves to protect and serve. If you asked Tommy if he was a hero, he'd give you that happy-go-lucky smile and tell you nonchalantly, "No, I was just doing my job."

Tommy's body was never recovered, the only remnant found was his FDNY helmet. He is with his brothers in Heaven, looking down on his family and—I thank God—on me.

Tommy, I met your family after I returned from Iraq. My wife and I visited them on Christmas Eve, 2006, to honor you and pay our respects. There were tears, reminiscences, and the knowledge that your spirit would always be present.

As we were getting ready to leave, Janet, Tommy's widow, presented me with his Mass Card, telling me to keep it with me wherever I went in the world, that Tommy would be my Guardian, Angel. I have. Tommy will always be with me, my Guardian Angel!

Tommy, this book is dedicated to your memory, your life and the lives of those who sacrificed themselves so that others may live.

Thomas Gambino, Jr.

Preface

We have all heard wonderful stories about our military working dogs (MWDs). There's Stubby the hero dog from World War I (WWI). And Jack, the Marine Scout dog, who served in the Pacific during World War II (WWII)—and the Korean War scout dogs.

Smokey, a Marine Scout dog, saved countless lives from Viet Cong ambushes. And most recently, Navy Special Warfare Dogs—Cairo and Conan—participated in the Ben Laden and Al Baghdadi raids.

Canines have served in combat roles since the beginning of recorded history. They performed as sentries guarding military facilities and as messengers braving not only hostile fire but also severe weather conditions. They've been used as pack animals delivering medical supplies. And since WWII, they are now performing as single- and dual-purpose explosive and narcotic detection dogs. They have been credited with saving thousands of lives.

The former Commander of U.S. Central Command, General David Petraeus, said, "The capability they (MWDs) bring to the fight cannot be replicated by man or machine. By all measures of performance, their yield outperforms any asset we have in our inventory. Our Army (and military) would be remiss if we failed to invest more in this incredibly valuable resource." [Reference 1]

The public is unaware of the daily fight to sustain this program. Most civilians don't see the struggle to obtain funding for procurement and training. They don't appreciate how antiquated our current methodologies have become. In fact, most citizens don't know how the system is structured or how it works. For example, I suspect that most will react with surprise on learning that the Air Force is responsible for the military dog program. They will, most likely, be alarmed to read that the Air Force, the Department of Defense (DoD) appointed Executive Agent (EA), acquires dogs solely from European breeders—and that source is dwindling.

The International Working Dog Breeders Association's (IWDBA) 2019 symposium "White Paper," well known within the international canine community, analyzed current trends in canine procurement. To appreciate the problem, we must first understand their basis. Dogs are best suited for military applications between the ages of eighteen months and two years of age—and be free from genetic diseases. Then they are evaluated for general obedience and, in the case of patrol dogs, for a defined level of aggressiveness.

The White Paper expressed concerns that within the next five years, dogs within the eighteen-to-twenty-four-month age group would no longer be available from European breeders. They reported that competing foreign governments were procuring the bulk of trainable dogs for up to $30,000. While the U.S. was restricted to a price cap of $5000. Unfortunately, the EA ignored this White Paper, and as a result, this situation is impacting military and law enforcement canine readiness.

Lack of resources and DoD priorities has negatively affected the health and welfare of the MWD force. Over forty percent of service kennels have major structural issues. DoD has Zero dollars for maintenance, and these facilities are not a priority for new military construction or major improvements. While the public only sees the heroics of our dogs and their handlers, behind the scenes our MWD superstars, especially our veterinarians, work miracles providing MWD health, safety, and welfare. However, Army Veterinarians are operating at minimal manning levels. And for those who are unaware, the Army provides Veterinary Services for all MWDs and many of our government law enforcement agencies. They are overburdened— not only with caring for military canines, but veterinarians are also responsible for food service inspections of our commissaries globally. And this reinforces the fact that there are no funds allocated for canine research and development (R&D). R&D is the responsibility of the Defense Health Agency (DHA). For years, multiple opportunities and projects have remained stagnated, sitting on the drawing board. Non-existent. Funding restrictions prohibit DHA from entering joint collaborations with Academic Institutions of Veterinary Science. Sadly, DHA remains in R&D dark ages, while academic institution research is fully funded, and available to compete for government grants. Will Veterinarians be capable of supporting future contingency operations?

We were lucky that COVID was not a zoonotic disease (transmittable from an animal to human). If it was, the spread of COVID not only from working dogs, but from other animals (pets, feral cats, strays), would have had the same devastating effect as the Plague that swept Europe in the sixteenth century. Let's not avoid the issue of chemical-biological decontamination. At this moment, there is no personal protective equipment (PPE) for working dogs or other animals. In a domestic emergency or act of terrorism, working animals would immediately become casualties, possibly eliminating the entire first responder detection capability.

As the former DoD Canine Program Manager, it was a daily battle to adequately serve the joint services. We identified critical military concerns and built international and domestic collaborations with our military, law enforcement and civilian canine partners. Even so, "canine" is a minuscule blip on the DoD radar screen. Compared to a ship, plane, tank, new barracks, and airfields, they are a low-cost line item. Only when military dogs become good human interest stories does Congress and the media give them airtime. When the stories fade, so does support and interest in the program.

Canines are so far down the totem pole that our voices are muffled, drowned out by chains of command with no understanding of the program—until the next terrorist attack, or the next natural disaster. Then the call for dogs will ring loud and clear.

This book presents the behind the glass door story of the MWD program. I'll show how the greatest team in the world battles uphill challenges in the Pentagon and the halls of Congress. Underfunded and mismanaged, a phenomenal team of canine heroes fights to keep our four-legged warriors trained and ready to meet our nation's challenges at home and abroad. I'll discuss how this low-tech capability remains the fatherless stepchild of the military. And I'll explore what will happen if we continue down this path.

Table of Contents

PREFACE .... *III*
CHAPTER 1 – HISTORY OF MILITARY WAR DOGS *3*
CHAPTER 2 – EXECUTIVE AGENT MISSION *10*
CHAPTER 3 – DOD PROGRAM MANAGER STAFFING *19*
CHAPTER 4 – CANINE TRAINING *23*
CHAPTER 5 – TRUE COLORS *27*
CHAPTER 6 – OPERATIONS IRAQI AND ENDURING FREEDOM *32*
CHAPTER 7 – CANINE FORCE FLOW *42*
CHAPTER 8 – MWD HUB MCGUIRE AFB – COMING HOME *48*
CHAPTER 9 – MWD HUB KUWAIT *50*
CHAPTER 10 – CONTRACT WORKING DOGS *55*
CHAPTER 11 – CADAVER DOGS *58*
CHAPTER 12 – FORCE MULTIPLIERS – TRADITIONAL TO SPECIAL *63*
CHAPTER 13 – TACTICAL EXPLOSIVE AND IMPROVISED EXPLOSIVE DETECTION DOGS *67*
CHAPTER 14 – U.S. SECRET SERVICE MISSIONS *84*
CHAPTER 15 – DEFENSE SUPPORT TO CIVIL AUTHORITIES *89*
CHAPTER 16 – DRUG LAB *92*
CHAPTER 17 – WORKING DOG MANAGEMENT SYSTEM *98*
CHAPTER 18 – INTER-AGENCY STORIES *101*
CHAPTER 19 – WHERE WE ARE HEADED *116*
CHAPTER 20 – CANINE LEADERSHIP TRAINING *118*
CHAPTER 21 – PROCUREMENT *121*
CHAPTER 22 – CAPABILITIES BASED ASSESSMENT *129*
CHAPTER 23 – R&D INTER-AGENCY COLLABORATIONS *131*
CHAPTER 24 – CANINE INTELLIGENCE *137*
CHAPTER 25 – STRATEGIC PLANNING *142*
CHAPTER 26 – INDEPENDENT ASSESSMENT *148*
CHAPTER 27 – STRATEGIC LESSONS LEARNED *151*
CHAPTER 28 – FORGOTTEN INFRASTRUCTURE *157*
CHAPTER 29 – NO SEAT AT THE TABLE *161*
CHAPTER 30 – RECOGNITION AND TRIBUTES *163*

CHAPTER 31 – EXEMPLARY INFRASTRUCTURE – ARMY STYLE *165*
CHAPTER 32 – GONE BUT NEVER FORGOTTEN *168*
CHAPTER 33 – SACRIFICE AND REMEMBRANCE *173*
CHAPTER 34 – AWARDS AND DECORATIONS *175*
CHAPTER 35 – REFLECTION - EQUALITY *182*
GLOSSARY *185*
REFERENCES *188*
ABOUT THE AUTHOR *189*

The Program — the Chess Board

Chapter 1 – History of Military War Dogs

Ancient Warfare

Since the beginning of recorded history, dogs have been used to support military combat operations. They comforted our wounded and detected explosives. They diverted criminals trying to inflict harm on innocent civilians. And in wartime, they protected military personnel by detecting improvised explosive devices. They've been invaluable, especially in Iraq and Afghanistan.

In murals as far back as the Ancient Egyptians, the fighting spirit of Egyptian war dogs are portrayed and commemorated. The emperor Hammurabi equipped soldiers with huge war dogs. In ancient Greece, the Corinthians used them as shoreline sentries to defend against Athenian amphibious assaults. Roman legions were the first to realize the need for medical care for these great mastiffs—and employed the first embedded military veterinarians.

The Romans were the first to look at canine capabilities strategically. They employed them as watchdogs, sheepdogs and hunting dogs. Hunting dogs had unique skills and were used to perform three specific tasks: hunting game for meals, tracking human and animal scent, and chasing and attacking enemy soldiers. Attila the Hun used packs of large dogs as sentries to prevent surprise attacks. During the medieval period, large war dogs, such as mastiffs, were clothed in chain mail and released to attack enemy horses. This function was a precursor to our modern tanks, in that, those animals were literally a mobile "armored" force. Napoleon, like the Romans, also deployed fighting dogs in front of his reserves in the early 19th Century.

The world continues to evolve. Throughout history, dogs were early warning tools. Now we know they can detect diseases as well. Their sense of smell cannot be duplicated by technology. As the history of the world continues to unfurl, our reliance on our four-legged warriors—ever vigilant, ever ready, proven in war and peace—continues to grow. They are truly man's best friend.

United States Seminole War

The first recorded use of dogs by the United States was during the Second Seminole War. The Army bought thirty-three Cuban-bred bloodhounds. These animals and their five handlers tracked Seminole Indians and the runaway slaves they were harboring.

United States Civil War

Officially, there was no organized MWD program during the Civil War. However, as both armies marched through the countryside, they adopted dogs as unit mascots and individual pets. There are two Civil War mascots of note. "Fan" provided moral support for her adopted regiment, the 88th New York. The 11th Pennsylvania Volunteers adopted "Sallie." She became so attached to the unit that she led the Regiment in the pass and review for President Lincoln after the Battle of Gettysburg. A cast replica of Sallie stands at the base of the granite monument to the 11th Pennsylvania Infantry at the Gettysburg National Military Park.

Sallie on the 11th Pennsylvania Infantry Monument at Gettysburg (Photo by J.K. Faulkner)

Spanish-American War

During the Spanish-American War, Teddy Roosevelt's "Rough Riders" used dogs as point scouts. They proved to be invaluable. Their use made it almost impossible to be ambushed. In Edwin Richardson's book [Reference 2], he describes Captain M.F. Steele's efforts to support a formal MWD program based on his experience in the Philippines. His recommendations concluded that, "dogs are the only scouts that can secure a small detachment against ambush on the trails through tropical jungles." In that environment, the foliage was so dense that flankers were out of the question. The leading man, at one or two hundred yards, was out of sight of the main party. However, their dogs' hunting instincts and olfaction (ability to smell) could detect insurgents anywhere they lurked.

Up until this period, dogs were acquired through ad hoc means, not sanctioned by the War Department or our allies' Ministries of Defense. "Dog people" unofficially procured and trained them for their own units. When conflicts ended, the functions—and the animals and their keepers—faded into the sunset. What was hastily stood up, was just as quickly stood down. And for a long time, the military functions performed by canines remained informal and undefined. For almost seventy years, successful working dog applications in the Philippines during the Spanish American War were forgotten.

World War I

Ironically, our British allies—capitalizing on Captain Steele's thesis on the value of military dogs—established a MWD program for WWI. The British, French and the International Red Cross used canines in four important capabilities—as messengers, sentries, and as ambulance and sanitary dogs. These highly trained animals saved countless lives by leading rescuers to wounded soldiers on the battlefield. They also delivered critical messages through enemy fire and across perilous terrain. The first use of dogs as a disease prevention tool was in WWI. As trench warfare was rife with rats, Jack Russell terriers helped eliminate the threat of soldiers being infected with disease (zoonotic, in this case) passed to them by these vermin.

Today, the U.S. Military has stringent selection criteria for canine candidates. Using multiple evaluations—as well as operational and veterinary screening protocols—it developed a "best breeds" list. A limited comparison of breeds produced a unique conclusion. Purebreds showed no advantage in performance over mixed breeds (Michael Lemish [Reference 3]). During WW I, there were no formal screening criteria to rely on. Bulldogs, retrievers, terriers, sheepdogs, and shepherds all saw service.

The French and British recognized canine intelligence and were the first to put it to work. The French initially used military working dogs from 1906 to 1914. However, with the onset of WWI, they reestablished a robust program. At the height of the war, the British ramped up thousands of dogs for the war effort too. Ultimately, the Germans used six thousand canines, the Italians three thousand and the French nine thousand.

Of all the nations embroiled in WWI, only the United States failed to see the strategic significance of canines as a critical asset, or "force multiplier." American commanders not only advocated against using dogs for most military applications, but they also discouraged their use in the U.S. military's leadership programs. Our senior military service colleges and DoD programs for the Senior Executive Service did not teach the value of war dogs.

And now? Nothing has changed. The lack of introduction and education on the basic functions of MWDs profoundly limit commanders' ability to integrate them into their operational planning. The first time Commanders are aware of dog team potential is when they arrive at their unit. They don't understand how to use them. This is because the DoD is ambivalent about educating leaders in what dogs bring to the war fight. They need to understand how these resources should be deployed and utilized. Most importantly, they need to appreciate that special veterinary care is essential to sustaining valuable canine detection and olfactory capabilities.

At the conclusion of WWI, there was no disposition plan. What do we do with the thousands of French, British, Italian, Russian dogs? Deemed as excess, the majority were destroyed, along with the termination of their working dog programs. In some cases, handlers were allowed to adopt their dogs…but that was the exception, not the rule. Only the United Kingdom (UK) retained a canine school. Realizing the effectiveness of positive operational capabilities, they strategically positioned their canine corps for another world war, which turned out to be only two decades away.

World War II

Dogs had proven themselves in combat. The need existed. How would the US War Department procure and train enough animals to support a global conflict? Robin Hutton [Reference 4] describes the collaborative effort between the American Kennel Club (AKC) and the War Department to create the "Dogs for Defense" program. Captivating the nation, Dogs for Defense encouraged private owners to donate their pets for wartime service. The national reaction was amazing.

Procurement for the Dogs for Defense program was totally domestic. We were self-reliant. Our cost-effective program eliminated paying third-party overseas breeders. Pet owners played a vital part in the selection process, as their pets were already socialized—and medically screened. Dogs for Defense was a national stakeholder program. Because of the numbers of dogs considered for donation, the War Department was able to select from multiple breeds. And by transitioning dogs from civilian to military use, they easily met the high demand for animals appropriate for security and detection missions, domestically and in the theaters of war.

Screening, accepting, and using dogs was tasked to the military services. The Army's Quartermaster Corp established "canine training depots" at Front Royal, Virginia; Fort Robinson, Nebraska; Camp Rimini, Montana; and San Carlos, California. The Marine Corps trained at New River, North Carolina. The Elkins Park Training Station in Pennsylvania prepared Coast Guard dogs. Others trained at Hilton Head, South Carolina. By 1943 through the end of the war in 1945, close to ten thousand dogs were accepted and used for the war effort.

The rapid growth in the MWD program was not fully funded by Congress. The government and AKC promoted a fundraising campaign known as, "Dogs for Dollars." This program raised thousands to offset logistical costs (kennels, dog food, i.e.). Corporate America rose to the occasion—Ralston Purina donated all the dog food for new canine recruits until they were shipped to training camps.

WWII provided a template for recognizing canine structure. MWDs were organized as units, under Tables of Organization and Equipment (TOE). The TOE documents the unit functions, the chain of command, manpower and resources. The British Army was the predecessor for today's canine organizational structure. They established a formal war dog program during WWI. While the other allied nations demobilized their canine programs, the British saw the strategic value—and while dramatically downsizing, they never eliminated the program during the lull from 1919 to 1939. Reducing resources and funding, they maintained minimal manning and a functional training school.

WWII opened doors for innovative and outside-the-box thinking. Almost twenty thousand dogs were donated. We sourced a global need almost overnight. After WWII, we eliminated domestic procurement. By the 1970s, the DoD had become reliant on overseas breeders from Europe. Now we must compete for canine resources. China, Russia and Saudi Arabia have the financial advantage, and are willing to pay five times more than what the DoD budgets for a trainable dog. A budget that has had minimal growth over the past twenty years! Lessons learned from WWII have been lost.

Korean War

At the start of the Korean War, the 26th Infantry Scout Dog Platoon trained at Fort Riley, Kansas. Their force structure was one hundred sentry dogs. They were immediately sent to Korea for combat patrols. On July 11, 1951, the War Dog Receiving and Holding Station was activated at Cameron Station, Alexandria, Virginia. This was where newly purchased dogs were processed and conditioned before being shipped to the Army Dog Training Center in Fort Carson, Colorado. On December 7, 1951, the responsibility for dog training in the continental United States (CONUS) was again transferred—this time to the Military Police Corps. And finally, early in 1952, the dog training center, along with the 26th Infantry Scout Dog Platoon, moved from Fort Riley, Kansas to Camp Carson, Colorado (later Fort Carson).

The outstanding results with the 26th Infantry Scout Dog Platoon led to plans for scout dog platoons for each division in Korea. However, only five platoons trained and shipped before the war ended. During the Korean War, members of the platoon received three Silver Stars, six Bronze Stars for Valor, and thirty-five Bronze Stars for meritorious service.

After peace negotiations ended the fighting on May 4, 1954, the War Dog Receiving and Holding Station at Cameron Station was placed in a stand-by status. Scout dogs not assigned to infantry divisions were retrained for sentry work to patrol the demilitarized zone that was established between North and South Korea.

The Army Dog Training Center continued at Fort Carson from 1954 to 1957. Then, there was a dramatic shift as most dogs being trained at Fort Carson went to the Air Force. In 1957, the Army center was deactivated…and the responsibility for training all dogs for military purposes was transferred to the Air Force.

To address this new responsibility, the Air Force established a permanent training center at Lackland Air Force Base (AFB), Texas. With the Vietnam mobilization in the 1960s, the Air Force was ill prepared to meet the challenges and numbers of dogs needed to support combat operations.

Army Dog Training Center, Fort Carson, Colorado February 1953

MWD Training- Lackland AFB, San Antonio, Texas late 1950s

Vietnam War

The United States Army Infantry Center, Fort Benning, Georgia, established Headquarters Detachment Scout Dog training with two attached platoons—the 26th Infantry Platoon (Scout Dog) and the 51st Infantry Platoon (Scout Dog). Together, they conducted training for dogs and handlers assigned as scout, combat tracker, or tunnel dog team duty in Vietnam. The Army Infantry School served as a Joint Service training course, training Army and Marine scout dog teams. The 60th Infantry Platoon became the third Scout Dog Platoon, with the added responsibility of training Army Mine/Tunnel Detection Dogs. These dogs were trained to detect the tunnel openings where enemy soldiers hid from American patrols. As the troop strength in Viet Nam approached a half a million the Army added a second canine training facility at Fort Gordon, Georgia.

The Army and Marine scout and tracker dogs supported search and destroy missions in the dense jungles. The Air Force and Navy sentry dogs were used strictly to provide perimeter security for static bases and ports, primarily Air Force bases in Viet Nam and across the globe. These dogs were trained specifically for patrol (bite work).

A total of 4900 MWDs served in Vietnam. The Army deployed twenty-six scout dog platoons and twenty-two combat tracker teams (platoon-sized elements). All four branches of the military used canines in Vietnam. Approximately ten thousand handlers served in Vietnam, the largest deployment of MWD teams in any of our nation's wars. It is estimated that these teams saved over ten thousand lives. The following is a breakdown of MWD teams by military branch that served in Vietnam: sixty-five percent - Army, twenty-six percent - Air Force, seven percent - Marine, and two percent - Navy.

The biggest travesty of the Viet Nam war was the MWD policy of not redeploying MWDs to the U.S. with their handlers. Unfounded concerns that dogs would spread zoonotic disease led to the policy that dogs remained in country. When we exited Vietnam, military working dogs were classified as "surplus equipment." The majority were euthanized, transferred to the South Vietnamese military and police, or just abandoned. Despite pleas from many handlers who were willing to pay for their dogs' flights home, the military wouldn't permit it. It is estimated that of the four thousand MWDs that served, fewer than two hundred made it back to the U.S.

But what the military realized from Viet Nam, was the need to maintain a permanent MWD capability in the DoD. In 1974, the DoD's assessment of which service should assume this responsibility resulted in the Air Force being designated as the Executive Agent and directed to establish a joint service MWD school. Today that responsibility is tasked to the 341st Training Readiness Squadron (TRS) at Lackland AFB. Their sole mission is to procure and train dogs for the military.

Operation Desert Storm

Operation Desert Shield/Storm deployed one hundred eighteen Joint Service MWD teams. As the war ended within one hundred hours, a large U.S. presence of MWDs was not required. In contrast to U.S. MWD involvement, the French deployed one thousand one hundred and seventeen MWD teams during the buildup in Operation Desert Shield.

Operations Iraqi and Enduring Freedom

At the height of Operations Iraqi and Enduring Freedom, the military had over six hundred MWDs and four hundred contract working dogs deployed in the Central Command (CENTCOM) theater of operations. The workhorses were explosive detection dog teams. Strategically, there was a huge operational shortfall. Dogs were in high demand. Every maneuver unit, Air Base, and forward operating base (FOB) requested MWD teams to detect explosives and IEDs (the insurgents' weapon of choice). Needed to provide base security and support counter-narcotics eradication missions in Afghanistan. Strategically, the decision was made to consolidate MWDs by capability. To that end, Army, Navy, and Air Force explosive and narcotic dogs trained to the same standards at Lackland. Deployment depended on the capability of the dog, not the uniforms of their handlers. Teams were jointly deployed to support priority of effort missions. The strategic mission was the consolidation of resources to achieve joint unity of effort. The new paradigm: Army maneuver units could be supported by Navy and Air Force explosive and narcotics detection dogs. The interoperability of resources was the new mantra for MWDs.

The new reality for all future conflicts— Joint Interoperability.

Chapter 2 – Executive Agent Mission

The DoD appoints Executive Agents to oversee, monitor and fund DoD operational programs. DoD Instruction 4000 directed the Air Force to assume this role for Military Working Dogs (MWDs). This mission was passed to the Air Force Office of Security Police in DoD Directive 5200.31E (7 Sep 1983) [Reference 9], where it remains. The Directive is general in nature and provides little if any accountability for program oversight. It has yet to be validated for consistency or efficacy. And most disconcerting, it does not answer the question, "does the program meet the basic question of public policy—the common good?" Those who have served in the program would reply with a resounding, "No!"

The main function of the Air Force is to fly, fight, and win…airpower anytime, anywhere, not train dogs. Air Force leaders, as well as taxpayers, understand that mission. In the public eye, that means all equipment and personnel expenses should be dedicated to airpower. The other military services, the Army, Navy, Marine Corps and Coast Guard have specific defined missions. The Air Force has no business…nor can they execute…a global strike canine program. That was proven tenfold during Operation Iraqi and Enduring Freedom. Canine should be under an independent Joint Organization funded and equipped to strategically execute a canine global strike mission.

DoD Directive 5200.31E states that EA responsibilities are limited to identifying requirements and resources necessary to provide dogs for the military services. It further states that the EA performs…but is not limited to…oversight of the MWD inventory, distribution, and disposition (retirement, adoptions). But it doesn't mention strategic or joint collaborative responsibility. Who then provides the oversight of the MWD enterprise to the Secretary of Defense? Apparently, no one. That critical flaw came to the MWD forefront when the U.S. Forces Afghanistan Commander (USFOR-A) placed an urgent request for additional explosive dogs. When that button was pushed, IEDs were causing more combat casualties than those from indirect fire from the Taliban. When the U.S. Forces Commander in Afghanistan requested two hundred dogs to defeat these threats, the Air Force response was…we can't help. They passed the torch to the Army and Marine Corps to procure and train their own dogs. Has anything changed since they said "we can't help? NO! Has the AF instituted global MWD strategic analysis? Did they return to the drawing board and commence planning for possible "surge requirements" to support multiple theaters of war? You know the answer…the result…when the next war comes and there is a surge for MWDs I truly believe the Air Force will be unprepared to fulfill their EA responsibilities and again pass the buck to the services…and troops will die needlessly.

The Executive Agent believes they have no place at the DoD canine readiness table. They just mope along at their own pace. But since WWI, the modern battlefield has been a Joint – Coalition collaboration. No one will stay in "their lane." To be successful, we'll need to develop the skills to work with partners and cross-level resources, mutually supporting our team of military, civilian, coalition and contract stakeholders. Iraq and Afghanistan identified the fact that dogs will be utilized based on capabilities and need, not by the handler's service uniform.

Where should the EA be engaged? The EA should assume additional responsibilities as the MWD advisor/force manager for the Joint Staff. The DoD should be directly involved as a partner, providing guidance on employment, capabilities, training, and deployments for the Combatant Commands. The EA should be providing a global MWD strategic footprint. During my tenure, I attempted to adapt my eight years of Joint – Coalition MWD operations experience in twenty-six countries to assess and develop a global strategic platform. For my efforts to try and move the AF past 1999 paradigms, I was told that I had gone "Native."

The EA and DoD MWD PM should be strategic, tackling MWD global strike capabilities. The EA needs to be the MWD lead for the DoD as part of the process, with strategic MWD thinking able to engage in these very basic questions.

   a. COCOM's strategic MWD footprint.
   b. Needs projections: What types/capabilities are required.
   c. In transit movements.
   d. Service fair share to meet the demand.
   e. MWD Organizational support: Conventional, Special Operations, Force Protection (Air Base, Ports, Installation)
   f. The MWD Chain of Command
   g. Identify Sources of MWDs (US, Coalition, Contract)
   h. Threats (environmental, toxins, chemical-biological, radiological)
   i. Pre-deployment training

This was quite evident with the EA's reluctance to inject themselves into Operations Iraqi and Enduring Freedom. The DoD MWD Program Manager was a key stakeholder in the MWD equation. How can the DoD execute an MWD program from behind a desk in San Antonio? How did the EA expect to have any concept of the focus of training, types of IEDs and tactics that they needed to address in training if they remained incognito? I invited the DoD MWD Program Manager multiple times to accompany me on our CENTCOM MWD assistance visits. The EA frowned at sending the DoD MWD Program Manager to the theater. It took some string pulling, but we finally got my dear friend and mentor, Bob, the DoD MWD PM into theater. He needed first-hand knowledge, this was a new war. Dogs were the only capability that could detect this new threat of homemade explosives. Every commander wanted to keep their troops safe. Where possible every "outside the wire" mission had a dog team. However, the Air Force remained complacent, safe inside the "Status Quo." Where I sat at CENTCOM, this was an all-hands-on deck war! The Air Force EA must have missed that memo.

For all intent and purposes, the DoD MWD Program Manager is the sole operational, tactical, and strategic manager for the Department of Defense. There is no MWD Program Manager/Advisor on staff of senior military commands. Not on the Joint Chiefs of Staff, or the geographic Combatant Commanders (Europe, Indo-Pacific, Northern or Southern Command). This is a systemic problem. Yet the DoD MWD PM is employed by the Air Force, the job evaluated by Air Force personnel, based on AF…not DoD/Joint…performance standards…and by a civilian personnel specialist in San Antonio with no MWD operational, tactical or strategic experience.

The other cog in the in the DoD policy chain is the Office of the Secretary of Defense (SECDEF). The SECDEF for Intelligence and Security is at fault. They too have not performed a program "needs assessment." Not since they have been in existence have they taken the initiative to send a representative to a Joint Service MWD Committee meeting. They provide paper oversight.

The Executive Agent provides guidance through Air Force Instruction 31-126 (Jan 2019) [Reference 10]. It provides program oversight to the subordinate military services. Loosely written, it reinforces Air Force control of the program. All regulations should be specific, but this one is not. Open to service interpretation. In practice, when the Air Force fails to meet its responsibilities, it passes their policy failures down to the services, relinquishing their DoD EA mission. The services forced to pick up the pieces. Thus, the services become scapegoats, having to internally Rob from Peter to Pay Paul. The

TEDD program is a shining example of Air Force program mismanagement. Incapable of meeting their responsibility to provide increased demands for MWDs for the warfighters in Afghanistan, they relinquished the mission to the Army and Marine Corps, directing them to temporarily fund and run their own programs. The repercussions to the EA… directed to revise the Air Force Instruction, minor policy corrections. Zero accountability. Will this become the standard EA practice during our next conflict?

Executive Agent Functionality

The EA sits in the Pentagon. It plays no part in the strategic employment and utilization of canine force multipliers in the global enterprise. The EA's purpose is the same as their predecessor, the Army Quartermaster Corps of WWII—procure and train dogs. In essence, the EA's mission is commodity management. The EA has no military function. Even when new policy modifications…as canine research and development, counter-narcotics, and possible enhancements of domestic procurement…were included in the 2020 DoD Directive 5200.31E, they were glanced over, treated as no more than a policy "tap dance." Nice work, but no program funding or additional personnel! Impact on program improvement equals zero. Baseless changes.

An assessment of the DoD mission clearly demonstrates the efficiencies, efficacy and cost benefits of a civilian contractor or academic veterinary institution (Auburn, Penn, Purdue, and Southern Illinois University) that could easily replace the purely administrative and training responsibilities of the 341st Training Readiness Squadron, providing high quality, high endurance, medically fit dogs. And use sound business practices to turn a "MWD Profit." This is especially evident as Auburn University entered into an agreement with the Department of Homeland Security and NYPD, to provide genetically screened, high-quality dogs through their fully funded breeding program. Within the past three years, the DoD National Security Agency has become a vested customer. All benefiting from this canine research and development Academic partnership. Auburn's program is light years ahead of the dismal lack of forward momentum by the EA.

The EA has not done a cost benefit analysis to determine program efficiency and effectiveness. In fact, a program review has never been performed. While there is a Capability Based Assessment (CBA) ongoing to identify MWD Program shortfalls, intended to gain a separate DoD MWD budget line item, the EA eliminated critical elements that were recommended for a full spectrum review. Originally, stakeholders recommended that the CBA be inclusive of costs for new kennels, research and development, DoD funded Secret Service missions and other critical MWD program needs. What was submitted to be evaluated… MWD procurement and training.

There was one exception in this decade of production failure. The 341st met their production goals in 2019. This was because the timeless concept of production of dual-purpose dogs was finally changed. After years of some unwritten law that German Shepherds and Belgian Malinois were the only breeds capable of serving as MWDs, common sense prevailed. The paradigm of dual-purpose dogs was broken with an analysis of MWD wartime missions in the middle east. After twenty years of MWD utilization, the facts were resoundingly clear. There was no requirement for an aggressive "bite" capable dog. Dogs trained to "Bite" were never used. During the twenty years of war in the middle east, Commanders requested single-purpose explosive detection dogs.

In May 2019, at the Joint Services MWD Committee meeting, the MWD decision makers, the Commander of the DoD Veterinary Hospital, Commander of the 341st Training Readiness Squadron, the Service MWD Program Managers unanimously voted to accept single purpose dogs within their

MWD ranks. What this did was open the flood gates, expanding availability to select from multiple breeds. Opening the door to start looking to reduce the reliance on overseas European breeders. Short haired pointers, Labradors, and Vizslas were several additional breeds the DoD could add to the selection pool.

The Marine Corps, and Navy immediately changed course, agreeing to start accepting single-purpose dogs, while the Army requirement remained dual purpose because of their wartime mission. The Air Force leaned towards a canine mix of dual- and single-purpose dogs. Finally allowing the services to decide their MWD make up. The customer, not EA, was now driving the production train.

In 2019, another major program improvement was implemented—the Fair Share Model. Eliminating the "Pooling" trained dog distribution process. Under the pooling concept, services who achieved or exceeded their Trained Dog Requirement (TDR) would have their excess dogs distributed to those services that fell short of achieving their annual production goal. A socialist concept of what was in the best interest of the Air Force and 341st. The newly introduced fair share model placed the onus of responsibility directly on the services to eliminate "pooling." It held each service accountable for providing sufficient manpower and internal analysis to ensure resources were in place to achieve their annual production objectives.

While the fair share model and expanding breed selection at the end of 2019 became the EA models, there was still a gap of over 250 trained dogs…and it had nothing to do with COVID. This was the culmination of the AF inability to meet annual service procurement quotas since 2013. What that gap is today is unknown, but any gap remains perilous to DoD canine readiness. In 2019 the revised Air Force Instruction unified all DoD canine assets under the Executive Agent. Prior to the 2019 revision, the four DoD civilian police agencies: The National Security Agency, Defense Intelligence Agency, the National Geospatial Agency, and Pentagon Force Protection Agency, operated under their own policy. There were no more hanging canine chads. That revision generated one DoD Canine Organizational oversight. That also allowed us to do a deep dive into their agency's procurement policies. How did they fund and purchase their dogs? They were at one hundred percent MWD authorizations. The AFI change also added additional explosive canine assets to the DoD inventory. Inclusion didn't detract from their program. We didn't direct them how to run their programs. Inclusion established a flow of information and added canine assets for possible deployments to assist with National Security Events. They also provided another capability lacking in the DoD, personnel screening dogs. We expanded our spectrum of operational capability, capturing their canine statistics. We also added their statistics in the annual report to Congress.

The mission of the Executive Agent is critical to our National Security. No other organization in the Department of Defense has the responsibility to provide a "living" force multiplier. The facts are clear. The history of the Air Force as the MWD Executive Agent is filled with flaws and program failures that should have sent up red star clusters to the Secretary of Defense. Their smoke and mirrors continue to bamboozle the SECDEF and House and Senate Armed Services Committees. The only rhyme or reason why the Air Force was made the EA, is that the Air Force has the largest inventory of dogs in the DoD. But it's clear that the EA is only a placeholder for policy and administrative oversight. The real mission of the EA is not only procurement and production. They need to be the strategic visionary for the SECDEF and Combatant Commands. The path of the EA is clouded. But time is running out. The clouds of war have never been darker across the globe than they are today. Will the EA be prepared to provide sufficient MWD force multipliers? Or will we repeat the mistakes of the past at the expense of lives needlessly lost?

Joint Service MWD Committee

I've talked about the Executive Agent; I've also briefly discussed the Trained Dog Requirement and the Fair Share Model. But you will want to know "Who is the governing body that sets the tone for "all things MWD." Who are the subject matter experts that bring their experience to the table? DoD 5400.31E established a subject matter expert panel to serve in an advisory role to the EA. The Joint Service Military Working Dog Committee (JSMWDC). Prior to 2019 the voting members were the Army, Navy, Air Force, and Marine Corps MWD Program Managers, the Commander 341st Training Readiness Squadron (TRS), Commander Holland Veterinary Hospital. In 2019 the additions of U.S. Special Operations and Central Command brought the needs of the warfighters to the program. The Committee meets every six months or as deemed necessary by the Chairman (DoD Program Manager). The agenda is determined by the committee members. As the Chairman, I leveraged unanimous agreements by the JSMWDC. I implemented the will of the Joint Services to drive canine readiness. If the JSMWDC brought an issue to the table and we voted on accepting the recommendation, it was "case closed." It was implemented. Several extremely groundbreaking initiatives were implemented by the Joint Services MWD Committee during my tenure:

1) FBI Memorandum of Agreement

> A no cost partnership with the FBI Office of Counter-Explosives. The FBI Explosive unit, by charter provided explosives training to law enforcement agencies nationally. They introduced the two most volatile terrorist explosives of choice—Triacetone Triperoxide (TATP) and Hexamethylene and Triperoxide Diamine (HMTD). Our military canines were not offered the same opportunity. TATP and HMTD training aids remain nonexistent. Opporture training was available, but many military canine teams remained unable to attend the training due to location and travel costs. For a short time, the military had collaborated with the ATF to imprint MWD on TATP and HMTD. The annual cost, $30,000, but less than a third of our canine teams were able to attend the ATF's training. The training was exceptional, and military teams received ATF certification.

> The FBI initiative would ensure MWD readiness for deployments overseas at military installations and supporting US Secret Service missions. The Memorandum of Agreement would allow the FBI to perform this no-cost mission to introduce all MWDs in the continental U.S., Hawaii, Puerto Rico, and Guam to TATP and HMTD. The MOA would ensure a joint inter-agency program to support DoD canine explosive detection annually—a huge step in DoD canine policy. Both agencies ensured that legal opines were obtained and i's dotted and t's crossed.

> When the MOA was ready for leaderships signature, and I completed the staff package I was informed that I had overstepped my authority. Asked why the DoD was the lead. Not by the EA but my non-canine chain. To my chagrin, they stated the typical AF position—the AF EA doesn't coordinate inter-agency training. Typical response. But as the DoD MWD PM, I represented all the services.

> The sole representative to the Executive Agent, my AF "handlers" were not in my chain of command, another story. This was a unified effort supporting the DoD. I assumed the responsibility and signed the MOA. We enacted a lifesaving, inter-agency collaboration that remains in effect today. Of course, the people who informed me that the DoD has no business of coordinating the MOA have since approved an extension, and of course the same people who chastised me, have since assumed credit, and praised the initiative.

2) Expanded JSMWDC Committee

In 2018, the DoD expanded participation in the semi-annual JSMWDC Committee meetings. Realizing that the DoD needed to break out of a self-imposed status quo we started building our team of stakeholders. This included Federal Law Enforcement Agencies (FBI, ATF, USSS, DHS, TSA) as well as the war-fighting combatant commands, U.S. Central Command and U.S. Special Operations Command. We modified the JSMWDC charter to include them as voting members. We then looked at global partners, including NATO, Israel, NYPD, FDNY, European Union, and Five Eyes (AUS, NZ). We expanded our outreach to academia who are fully funded and conduct canine research, including Veterinary researchers from Auburn, Purdue, and Southern Illinois University. We also brought on the American Kennel Club as a strategic partner. Their inclusion opened doors to help pursue a more robust domestic breeding program. We leaped from the frying pan into a global DoD canine stakeholder enterprise. One team, one fight…or so I thought.

In less than two years, we changed the façade of the program. We performed what the EA had failed to do—a Program Review. We analyzed current policies and protocols. Concurrently, we assessed future threats from an operational, tactical, and strategic level. We brought together a collaborative team of subject-matter experts from multiple stakeholder agencies and academic institutions, sharing information, and establishing a diverse and inclusive canine enterprise.

This was based on lessons learned from my eight years executing a wartime canine program. And it was—for the most part—transference. I was truly blessed. As we began rebuilding the program, I relied on the team. I was fortunate to have served with most of the service program managers during their many deployments. Their intuitive analysis and experience were instrumental to frame the new dawn. This organizational inter-agency collaboration—regardless of the support and institutional push back from the EA—was immeasurable. Success remained a win-win, but it was elusive to the EA.

For this rebuilding effort, for creating a global integrated interoperable enterprise, the forward momentum was put in reverse by my leadership. My so-called branch chief remained focused on capitalizing on my shortcomings with my inability to navigate the Tasker Management Tool (TMT) system. My efforts categorized as "Mission Creep." Unfortunately, the "mission creep" was not mine, it was Air Force minions who were self-serving, duly impressed with their positions. The DoD program was on track, we had unified the program, forging changes necessary to achieve canine readiness.

During my tenure as the DoD PM, I conducted five Joint Service Meetings. But, for my entire time with the AF, I was never allowed to brief the EA. Any joint canine issue had to be presented through an AF LTC—my branch chief—who had never had a canine command or ever served in a joint assignment. He was more concerned about appearances than the importance of the canine mission. I remained sequestered to my cubicle. Now this same former 05, the Air Force crony, took over my role, continuing to focus on AF only mission—true hypocrisy. Not one JSMWDC after action report ever made it to the EA. They ignored the recommendations brought by Joint Services. Regardless, the Joint Service MWD Committee persevered, opening lines of communication. We quickly changed the "stuck in the mud" meetings of having the services present their program briefings. We were tired of listening to AF excuses about their inability to meet their annual production numbers (TDR). We needed a meeting agenda that allowed us to establish action items in the best interests of the services—and then we needed to follow through on them. We reversed course. We identified problems and set the course for success. We began developing a five-year plan to ensure canine readiness. It included analyzing canine intelligence, looking at future challenges, global and domestic. We identified issues, discussed, voted and implemented recommendations. We were pro-active.

JSMWDC Research and Development Committee

The Joint Service MWD Committee received numerous presentations from outside organizations on the great work that was being conducted in academia and the DHS. We sat and listened, agreeing that the R&D for MWDs without funding was only a glint in our eyes. For too long the JSMWDC meetings were bitch forums with no action taken on Joint Service issues, especially recommendations on R&D.

When I was appointed as the Chair of the JSMWDC, we framed a working group to review the joint service committee charter. The JSMWDC was tasked with the responsibility to direct MWD research and development. We just didn't have the resources to drive that train. The veterinary community, now under the Defense Health Agency (DHA) who were the DoD veterinary subject matter experts were not included in the membership charter. But it was clear as a bell that this needed to be rectified.

For several months, Dr. (LTC) Jay Coisman—my dear DHA counterpart and friend—and I worked diligently to frame the language of what was to become the JSMWDC research and development sub-committee charter. We drafted, and re-drafted, going back and forth with stakeholders on the draft revisions. The Joint Service MWD Committee was operational, how do you train and employ dogs. The R&D sub-committees' primary mission would be a full-fledged partner and advisor for MWD R&D. They would bring together the veterinary community, MWD, and external stakeholders, in a formal forum to present R&D issues. They would provide guidance for inclusion of R&D in DoD policy.

We also realized that we needed to jump start this "New Dawn." With the Defense Health Agency taking the lead, it was the perfect time to establish a DoD MWD Canine R&D forum. The efforts of the DHA team went above and beyond. They rolled up their sleeves and pulled the entire forum together. We collaborated, bringing in our brethren from the national canine community. We defined DoD initiatives to project the MWD Research and Development enterprise.

In March 2019, that dream came true, and we conducted the First DoD MWD Research and Development Symposium. The turnout was amazing, I was in awe. Over fifty people from the full spectrum of the canine community attended. Members from Academia, DHA, FBI, CENTCOM, SOCOM, DoD 341st Training Readiness Squadron, DoD Vet Hospital, Army Futures Command, U.S. Secret Service, all the military services. We provided an open forum to focus on the direction to "jump start" DoD Canine R&D. Capitalizing…or the hope was to capitalize…on research, integrating all partners in the canine enterprise. This forum was groundbreaking, finally forcing R&D to the forefront as an integral element of the DoD Canine program.

Immediately, we identified the number one shortfall in the program. R&D was a paper program. The allocation for DoD funding for R&D…zero. But this forum was the catalyst that started the conversations to bring forth the critical need to tackle and pursue a course of action to change our current stagnation. It became a way ahead to overcome the lack of AF interest and funding.

How could the owner of the largest fleet of animals adjust to a changing global environment, improve canine conditioning, longevity, and performance—without R&D? But what was most notable…that this groundbreaking initiative, a new page, was not attended by the Executive Agent. The level of importance of the MWD program is consistent with the EA's overall support and interest.

That led to the current efforts to initiate the Capability Based Assessment. This was a cradle-to-grave analysis of the program, justifying a separate canine program appropriations line item of an accounting.

We identified and determined the priority threat to military canines …the same priority voiced by the entire community of stakeholders. We emphasized the total lack of decontamination training and protective equipment for canines exposed to chemical and biological agents. Out of sight…out of mind!

We did not encounter this during Operations Iraqi and Enduring Freedom, it became a non-issue, ergo a non-issue. However, with the introduction of Ebola, Anthrax, Hoof and Mouth disease, and COVID…a litany of canine threats need immediate protective equipment for dogs. The possibility that we could lose the entire canine detection force and experience a pandemic worse than the plague of the 1500s rose to the top of our priority list.

This forum was going to lead the way for R&D. Thankfully Dr. Erin Perry from Southern Illinois University, Dr. Eileen Jenkins, Dr. Coisman from DHA and our futures command team of Ms. Jenna Gadberry and Dr. Michele Maughan had already identified chem-bio protection as a priority of effort for R&D for chem-bio threats. Dr. Perry, who I believe is the nation's subject matter expert on canine chem-bio decontamination, had completed extensive research on chem-bio mitigation protocols. This forum, thank goodness, pulled together this tremendous team of stakeholders, helping to take us out of the stone age, waiting for disaster to happen.

Dr. Cindy Otto, from UPenn, did an in-depth article on the effects of fentanyl on dogs. She was also a leader in the movement to have civilian law enforcement officer canine teams issued Narcan. As we started pushing joint service military Drug Dog teams to the southern border supporting Joint Task Force North in 2017, they deployed unprotected, without Narcan. Military teams had to rely on CBP officers who accompanied MWDs on searches to make sure that if they were exposed to fentanyl the CBP officer could administer Narcan to MWDs.

The EA at the time felt that fentanyl was not a threat to military working dogs, the reasoning…military bases have not experienced an issue with fentanyl ergo it's not a threat. Another non-canine person making a call without any clue. Preparation and being proactive to protect our canines, why bother. Yet we knew that dogs exposed to fentanyl without the reverse agent of Narcan would be as devastating to the canine community as 9/11 and Pearl Harbor. We knew the threat was there, we knew that it was only a matter of time.

This is where common sense and the welfare of our canine heroes usurps an unaware chain of command and short sightedness of senior leaders. I brought this up as a discussion point in the symposium. Our civilian counterparts informed us. that their canine handlers were issued Narcan, one for the handler one for the dog. In a side bar with Dr. (Col) Andy McGraw the Commander of the DoD Vet Hospital the DHA team concurred that this issue needed to be addressed. Dr. McGraw and the medical team drew up a policy memorandum on the spot, providing guidance for issuing Narcan to dog handlers being deployed to areas considered high risk for fentanyl exposure. Common sense prevailed. We were able to provide the necessary tool to protect our MWD teams operating in a narcotic/opioid rich environment. We tactically bypassed the roadblock to get the mission accomplished.

Briefings continued from the AKC whose outreach and integration with the DoD could have solved the gap for procurement. Dr. Paul Waggoner from Auburn University discussed how their breeding program used behavioral and environmental distractors as part of the universities ongoing canine research. Introducing these distractors immediately improved the life cycle and performance of their single-purpose explosive dogs. We moved forward concluding on day three with concurrence that this symposium needed to be an annual event. The DoD has maintained that momentum, with the underlying knowledge that funding for any DoD R&D initiative remains a bridge too far.

Codification

In April 2019, we expanded our venue, selecting Auburn University as our meeting site for our semi-annual conference. Prior to April, every joint service meeting had been held either at Lackland or the five-sided maze in the Pentagon. The rubber was going to meet the road, as the voting members of the JSMWDC would be voting to modify the charter that had not been modified since establishing the EA as the Executive Agent in 1983.

As the topic to vote on the R&D sub-committee was introduced in the Auburn classroom, I looked around and was pleased at what I saw. The FBI, Special Operations Command (SOCOM), Army Futures Command, National Security Agency Police, the goal of having a cross section of stakeholders was expanding for the good of the program. In a unanimous vote, the JSMWDC approved establishing the Joint Service MWD R&D subcommittee. Our operational and veterinary stakeholders were going to be on an equal footing to be a unified advisory team for the EA, Combatant Commands, and the services. The committee was diverse and flexible, able to tailor projects with partners from members of the team outside the military community. The right partners with the right expertise. We added another pro-active layer to our DoD MWD family.

Unfortunately, all our hard work was stonewalled. I was prohibited from discussing any Joint Services issues with the EA. Only "real AF" officers were able to engage the EA. What was presented was no more than a bait and switch game feeding the general exactly what they wanted him to hear. It avoided any discussions on readiness or MWD strategy.

The Air Force was not big on the facts and the realities of the truth. I was not part of the embedded Air Force Culture. I represented my colleagues.

I stood true to my word and presented the truth and stood by my integrity. Being true to oneself and maintaining integrity was something General Powell reflected on, *"I have not allowed myself to be coerced...to provide very, very cheap (solutions) that look neat but won't accomplish the intended purpose."*

Chapter 3 – DoD Program Manager Staffing

Until I assumed the duties of the DoD PM, the position was located at Lackland AFB, San Antonio. From 2001 – 2007, the task organization consisted of the DoD PM, a deputy PM, an Admin NCO and two Airman operating the USSS mission tasking cell. Five people to support the global mission. After the retirement of Mr. Bob Dameworth, my friend and mentor, the DoD PM alignment shifted. The incoming Program Manager, retired Colonel Doug Miller, reduced the staff to an Air Force Master Sergeant as his deputy. The USSS cell was pulled and reassigned to a non-canine employee. In true Air Force forward leaning leadership, the civilian assigned oversight of the USSS tasking cell never had any contact with the USSS Technical Support Division. He never ventured outside San Antonio. He never spoke or attempted to establish a command relationship with the USSS. This was a point of contention during my tenure—and a constant battle with my division leadership. I maintained that this was my job. I had operational canine experience, and directed MWD dignitary missions protecting POTUS, VPOTUS and SECSTATE.

But the AFSF needed justification to retain and protect another Air Force retiree, regardless of the promises by the EA that the Secret Service Tasking cell would be moved to Secret Service Headquarters in DC. I voiced my opinion that effectiveness and efficacy to execute USSS routine missions and National Security Special Events in the best interest of the DoD required the tasking cell to be relocated to D.C. Again, I was told by my Division Chief that the Air Force Security Forces Center (AFSFC) at Lackland AFB would retain the cell. True to my job description I bypassed the AF blockade. I intervened, provided guidance, and made mission decisions regardless of the manpower maneuvering. The tasking cell walked a tightrope.

This was a direct violation of the DoD Program Manager job description. My job description was a document that was constantly manipulated when the Air Force felt it convenient. However, I pointed out that one of the job's many critical tasks was the sole point of contact for Secret Service dignitary protection missions. The document also defined my specific duties of inter-agency collaborations with the FBI and Department of Homeland Security. I was forced to bypass the AFSF Supervisor and deal directly with the tasking cell. I changed the perception of negativity and mistrust by establishing a monthly "Coffee" with the USSS Technical Support Division. A face-to-face partnership with our TSD brethren. When I mentioned the lines of communication that I had been directed to follow and mentioned the paper supervisor at Lackland, I was met with blank stares. They had no clue who he was.

I had the oversight for the four service program managers and four secret service tasking cell airmen. We changed paradigms, we did what should have been done a long time ago—work together.

The team pushed and I embraced being "rogue". As Gen. Powell stated, no one wanted to hear, *"Untidy Truth is better than smooth lies."* At CENTCOM, it was expected that we would provide the unvarnished truth. Without it, Senior Leaders would be unable to project the tools for success. But the truths that I was entrusted to bring to the EA were manipulated, replaced with what General Powell refers to as smooth lies.

My function was minimalized. "Shut up and color" was the flavor of this work environment. There was no understanding of "interoperability" or "joint service" operations. Many…especially the senior Air Force civilian leaders who had never deployed…relied on how the MWD program functioned in the 1990s. It was as if no one knew a war had ever happened. There was no desire to understand or listen to stakeholders who had deployed to the theaters of operation and no *desire* to implement lessons learned. I felt as if I were in the twilight zone. General Powell's words resonated as I "Soldiered On."

In retrospect, the military services are spot on with their program manning. The Army, Navy, Air Force and Marine Corps have MWD Program Managers that all have deputies. Even when the position was located at Lackland, the DoD program had at one time or the other, one or two Air Force canine deputies. That disappeared when the position was moved to the Pentagon. As the sole person having to absorb all previous "team" functions, my work schedule as a civilian exceeded twelve-fourteen hours daily, including weekends. I was on call and had no choice. The mission needed to move forward. Yet seeking compensation was out of the question. Air Force non-canine senior leaders told me I should budget my time. The concept of operation precipitating the fact that they had failed to meet procurement goals.

We place such emphasis on "Countering Improvised Explosives," but forget that the only capability known to man to positively identify explosives is a dog.

The same issue existed when the DoD stood up the "Joint Improvised Explosives Defeat the Device Organization" or JIEDDO. Established in February 2006, their mission, supported by millions of dollars, was to develop counter-IED mitigation equipment. However, a military working dog staff position— was not considered essential JIEDDO staff. Same story. All the technology in the world cannot replace the olfactory capabilities of a dog. Yet JIEDDO had an open checkbook given free rein to develop "wiz bang" counter IED technologies. Many prototypes were funded and never left the drawing board. And not one had the mobility and versatility to maneuver on the battlefield and urban terrain. If the living four-legged counter IED tool had been brought into the equation, they would have discovered that the dog doesn't need a battery or power source. It can be dropped from an aircraft and hit the ground running. It's less costly than high-tech sensors...and it needs less maintenance.

The depth of this program extends far beyond what can be accomplished by one person. I cultivated and worked to establish inter-agency relationships. I exploited years of joint service, law enforcement and coalition partnerships. I executed a program at my own personal expense (personal phone) and outside the boundaries of an eight-hour duty day. The tasks below are pulled directly from my job description. The international, inter-agency credibility, strategic knowledge, and theoretical design of Public Administration Policy had no value to the organization. Listed below is what I was required to accomplish, extrapolated directly from the job description.

a. The primary purpose of this position is: To function as Program Manager over Military Working Dog Program (MWD) and Department of Defense Counter Narcotics Program (MWD Support). This position has been delegated *complete authority over all Joint Service explosive detector dog teams which deploy in support of dignitary protection missions for the U.S. President, Vice President, First Lady, Secretary of State, Foreign Heads of State, and Foreign Dignitaries.* This averages 700-750 personnel supporting more than 300 missions per year worldwide.

b. Organizational Goals and Objectives: Effective worldwide management of the DoD MWD and DoD Counter Narcotics programs, as well as joint service support to non-DoD law enforcement agencies. Develop and implement effective joint service force protection initiatives:

—*Oversees and manages the Department of Defense (DoD) Military Working Dog Program.*

CORE RESPONSIBILITIES

- The DoD MWD Program Manager provides *single point* Executive Agency management over the DoD MWD program (includes Army, Navy, Air Force, and Marine Corps). Additionally, incumbent provides operational program management direction to the United States Secret Service, Department of State, the FBI, TSA, and the DHS concerning protection of the

President, Vice President and First Lady of the United States, foreign heads of state, foreign dignitaries and other dignitary protection issues mandated for DoD MWD support.
- Chairs the Joint Service Military Working Dog Committee and the Joint Service Trained Dog Requirements Committee. As DoD Executive Agency Program Manager and senior subject matter expert, has direct and recurring interface with senior military leaders of allied nation military forces by assisting with and affecting issues pertaining to MWD training, procurement, and combat operations. As DoD MWD program manager, takes lead as technical expert in interagency study groups in resolving problems in existing security systems.
- Develops innovative ways and procedures for solving problems not susceptible to regular or accepted security methods (i.e., in support of MWD, finding safer and deployable means in locating drugs/explosives).
- Resolves major conflicts regarding MWD policy and program objectives. *Advise top level agency security personnel on new developments and advances in security techniques. Recommends new methods or ways for enhancing the efficiency of security systems through modifications and applications. Evaluates and makes recommendations concerning overall plans and proposals for interagency security projects.* Implements national level guidance in agency standards, guidelines, or policies.
- *Provides oversight to HQ US. Army Veterinary Command Research and Development (R&D) and Special Operations programs.* Authors and maintains a MWD Requirements document recording essential research questions on MWDs. Seeks funding and manages execution of funds, as required, from Counter Terrorism Technical Support Office and other DoD organizations supporting R&D and other special projects. Solicits research questions and reports findings from MWD research, including NATO partners, to the Joint Service Military Working Dog Committee.

DUTY 4: Ten Percent Critical

Initiates, develops, authors, edits, updates, and interprets DoD, USA, USN, USMC, and USAF directives, instructions, and handbooks as they apply to the DoD MWD program. Develops and executes policy pertaining to the DoD MWD Program. Provides or arranges for MWD training and implementation directives as requested and needed. When developing directives/handbooks, provides solutions to security problems and questions and implement their use. Ensures national level guidance in agency standards, guidelines, or polices regarding MWD are enforced.

DUTY 5: Ten Percent Critical

Maintains currency on worldwide terrorism issues as they apply to explosives and Improvised Explosive Device (IED) technology. Applies this information to ensure continued effectiveness and process improvement to the explosive detector dog (EDD) federal dignitary protection program and protection of the President and Vice President of the United States and the global war on terrorism.

Rand Study

The Air Force contracted the Rand Corporation to do an independent manpower report in 2014. It was titled, "Air Force Manpower Requirements and Component Mix—A Focus on Agile Combat Support." They performed a quantitative study to assess total Air Force manning for all Air Force missions—Active Air Force, Air Force Reserve, Air National Guard, Civilian and Contractor Work force. Rand explored the qualitative side of personnel requirements. These included specialty qualification standards and other related processes. Further, they specified other required competencies of various workforces. There were no analysis of Air Force MWD – DoD MWD Program requirements.

The Rand study was based on Air Force conventional organizational manning, not a full spectrum analysis that included wartime expansion. As a result of the need for increases in military manpower, Congress lifted the personnel caps directed in the annual National Defense Authorization Act. This is standard practice, allowing growth of the services and Combatant Commands to meet the varying demands between peace time and wartime manning levels. Each of the Combatant Commands maintain their wartime augmentation requirements, based on skill sets, ranks, by service.

When we went to war in Iraq and Afghanistan, CENTCOM pushed the Joint Staff to fill wartime requirements, and the Joint Staff approved this growth directing the services to provide the additional manpower. Augmentees from Active Duty, the Reserves, and recalled retirees were mobilized to fill these positions. Rand had access to multiple unclassified lessons learned and after-action reports, that identified these wartime requirements. Why was the report limited to internal Air Force manning?

The report was flawed. Using Air Force manpower to fill Combatant Command Joint billets was a critical assumption. In the event of contingency operations, the five COCOMS (CENTCOM, TRANSCOM, SOCOM, EUCOM, INDO-PACOM) plus up their staff to augment their Joint Staffs in their respective Headquarters. Rand failed to include an analysis for joint force operations support of combatant commands. For one, The Combined Air Operations Center (CAOC) is the Air Operational element in a Joint service coalition that maintains command and control for unified air operations in a wartime theater. The CAOC, staffed primarily by U.S. Air Force augmentees, provides global strike capabilities. Whichever geographic region in the world, the respective COCOM will stand up a CAOC.

The report also makes no mention of the need to cross-level personnel. Yet by 2014, Air Force MWDs were being cross leveled to support Army units and more than thirty percent of Air Force MWDs were being assigned to support Army Brigade Combat Teams. The Air Force Security forces also performed a non-Air Force skill set…Army Detainee operations. At one point in Iraq, more than half of the units performing detainee operations were Air Force Security Forces units. The Air Force also joined the Joint Training Community, executing what was originally an Army mission, forming Police Mobile Training Teams (PMTTs). These assumptions were not included in the Rand report. But the lessons learned from what we experienced in CENTCOM are the same assumptions that the services and COCOM are using as planning factors in War plans. We learned that we fight based on the capability needed, not the color of the uniform. To believe that the Air Force, and particularly the Air Force Security Forces mission will be doctrinal Air Force is unrealistic. The Rand Report does not provide systemic analysis of supporting non-traditional Combatant Command global contingency operations. The analysis, based on misinformation, provided the Air Force Chief of Staff with a false quantitative narrative. What Rand should have been asked to do was a full global strike assessment inclusive of internal, contingency, and Defense Support to Civil Authorities missions.

Chapter 4 – Canine Training

The EA, through the 341st Training Readiness Squadron, provides basically trained dogs for the Army, Navy, Air Force and Marine Corps. Up until 2009, the 341st provided dual-purpose (bite detection) trained Patrol Explosive Detector Dogs (PEDDs) and Patrol Drug Detection Dogs (PDDDs). In 2009, the need to support mitigation of Improvised Explosive Detection Devices in Iraq and Afghanistan for the Army and Marine Corps required a new canine capability to save lives—an "off leash" detection dog that would provide a safe "standoff distance" for troops, while also providing an extended search area of up to 75 meters, the Specialized Search Dog (SSD) course. The other special program, the Combat Tracking Dog (CTD) course, provided off leash dogs specially trained to track explosive scent for long distances from the site of an explosion back to the source of where the explosive was built.

The 341st adopted these courses as Joint DoD courses of instruction. The bill payer…the services. The 341st only provided dogs and training space at Lackland. The Marines and Army had to provide the instructors. These programs proved extremely successful. The SSD and CTD programs terminated sometime between 2015-2017, but SSD teams remained in service until they completed their working life cycle. These programs demonstrated the need of a single purpose dog platform.

The 341st also has the responsibility for procuring Mine Detection Dogs (MDDs) for the Army. This program, originally introduced in WWII, requires dogs that the Army Engineer School trains to do what is called "Lanes" searches—methodical searches to detect buried mines/explosives at least twelve inches underground. The 341st does not have oversight of or approves the MDD training program. However, they do take credit for procuring MDDs for the army blindly, in the Annual Report to Congress. That seems odd as all MWD training programs are supposedly approved by the DoD. The other question that the EA should be asking, is why haven't Army MDDs ever been utilized in an operational mission? Why is the DoD continuing to pay for a program never utilized in its history? DoD program analysis is a systemic problem.

What the public may not be aware of, is that the DoD is also responsible for procuring dogs for the Transportation Security Agency (TSA). Through a memorandum of agreement, the DoD acquires training dogs for the TSA annually. Procuring dogs through a reimbursable cost sharing agreement. I always wondered why the DoD has the responsibility for procuring dogs for the Department of Homeland Security. The DoD has no oversight, training responsibility or accountability of the TSA dogs. The question is why? What's the Quid Pro Quo? Homeland Security has an open check book. The DoD has limited funds for procurement. There is am imbalance, that can only be fixed when the EA performs a full spectrum program assessment.

There is another operational shortfall that is affecting military canine readiness. Prior to 2009, the 341st employed civilian trainers, proven subject matter experts, to train dogs for the military services. From the inception of the Air Force as the EA until that point, the "Trained Dog Requirement (TDR)" was always achieved. Without exception! These civilian trainers provided continuity and expertise from their past canine careers as handlers. They were the cream of the canine training crop. Continuity and stability in training protocols was maintained and the military working dog handlers were where they should be, in their units, preparing for deployments and executing missions.

However, in 2009, the Air Force eliminated civilian trainers as an Air Force cost-savings move, imposing a reduction in canine readiness on the services. The concept devised by the Air Force to fund the basic handler's course was to direct trainers be levied on the services. An Air Force, not a Joint

Services decision. The JSMWDC should have been consulted. In essence, the Air Force attained free military manpower while saving millions.

This cost savings measure for the Air Force was a huge detriment to the other services, who had to break up certified combat ready MWD teams for three years. MWD Teams that should have been on rotation to deploy to Iraq and Afghanistan! Unlike civilian trainers, MWD handlers were not qualified trainers. Each had a different level of training expertise which meant each had a different level of productivity. This situation affected service productivity since unqualified trainers could not work dogs until they were certified by the Air Force. As a result, the program lost continuity.

What was the metric for trainers to produce dogs? It was a standard formula called the trained dog ratio. Based on historical data an anticipated benchmark of time and effort to train a dog was assessed. For example, statistical data was accumulated, and based on the average, a formula was determined. One trainer = time and effort = two trained dogs in an eight-week period (for explanation purposes only). Times the annual numbers of dogs required = numbers of service trainers. But the change failed to take into consideration that military trainers would be selected with different levels of training expertise, not every trainer was equal. A variable not included in the statistical equation. Statistics can't replace experience. This disparity instantly created a gap in annual productivity. Of course, the Air Force did not see that as their problem. The blame and lack of production rested on the services providing quality trainers. Air Force plausible deniability!

However, the Air Force, the 341st and 37th Training Group, still retain the Executive Agency, regardless of a decade plus inability to meet service production rates and its failure to support Combat Commanders' strategic needs. The program remains minimally funded by the "Air Education Training Command," and the gap of trained dogs continues to grow. These facts demonstrate that the Air Force does not have the proficiency, efficacy, and efficiency necessary to meet the mission of the DoD.

How could this be fixed? Could a contractor run this course…Absolutely! Returning efficacy and providing high quality dogs.

What grants the 341st the authority to train dogs for the military services? A convoluted Memorandum of Agreement (MOA) directed by Air Force Instruction 31-126 between the EA and the services. This MOA replaced the Inter-Service Training Review Organization (ITRO) and the Joint Service program directive. Rescinded in Nov 2003, giving the Air Force complete autonomy of the MWD training program.

Under ITRO, the Department of Defense funded the course, with the services projecting requirements, resources, training doctrine and policy. Under ITRO, the services had an equal say on course content and service handlers were not pulled from operational units to serve as trainers. The Air Force would be stripped of their single service power to run the school solely in their own best interests.

This battle has been a priority for the Joint Services Military Working Dog Committee and has been going on for years. However, the Executive Agent refuses to relinquish control, circling the Air Force wagons and sacrificing MWD readiness. With fentanyl now a global threat and without equipment to protect canines from chemical-biological agents, we've learned nothing from twenty years of combat operations. The only time Air Force Senior leaders engage with MWDs is when the media and Congress pressures them. They did cave when TEDD handlers demanded to adopt "Their Dogs," directing changes to the Air Force Instruction. Providing guidance on the process for non-traditional handlers to adopt an available dog. In 2018 Congress mandated that dogs adopted from overseas would be returned to the U.S. at government expense. However, Congress provided zero dollars in service appropriations to pay the airfare back to the U.S. The going airfare approximately $10K. Where was that money coming from? Existing service canine operational budgets! Congress threw the MWD program under the bus to appease constituents. Did the EA fight the issue, no, they allowed poor legislation to place a financial strain on the services. It had a tremendous effect on canine readiness, all because of what happens behind the glass door.

Now let's delve into the Memorandum of Agreement (MOA). The MOA provided the joint service concurrence for program funding, personnel (trainers) administration, program policies, curriculum reviews, Advisory Group/Program Board (which was non-existent until 2019), student administration, et al. It also directed that the course be accredited under the Air Force Community College, providing civilian education credits towards an associate degree.

Drafted and staffed through the Joint Service MWD committee, this MOA required approval from each of the services' training commands. Each service went through their own process to attain their separate General or Flag Officer approval. This staffing process took a year to 18 months.

Once signed, the services agreed to the terms dictated by the Air Force EA and Air Education Training Command. And prior to 2019, the services had no input. All course modifications were executed by an Air Force *only* STRT (Specialty Training Requirements Team) review panel. That is, until 2019, when we breached that barrier and the DoD PM and other services demanded participation in the STRT process. This change generated the methodology of service inclusion. This new found inclusion allowed course modification of what is now known as a basic level III canine. The inclusion in the course review in STRT streamlined the course and eliminated redundancy. We breached the Air Force barrier of exclusion by re-establishing the correct balance of Joint Service participation, after being shut out for more than a decade.

What made no sense was the period between MOA reviews. Nine years. When I came on board, we'd hit that nine-year mark. The MOA should have been reviewed every three years. And because it was on a nine-year revision cycle, it was out of date and in a cycle of stagnation. The MOA was antiquated. We had been engaged in twenty years of war. What was missing…guidance of what the ramifications would be if the Air Force was unable to meet the services trained dog requirements. The Air Force rallied behind their MOA to reduce the challenge for changes requested by the services. In essence, the Joint Service MWD Committee had minimal input in setting the stage to modify the 341st total control of the canine training program. An anchor around their necks.

Complicating the revision was the new exclusively well-functioning "tasker management tool." This new technology was supposed to revolutionize how action items, referred to as taskers, would be effectively and efficiently processed. A new streamlined process! However, after the Air Force spent millions to have this technology developed, the system could not interface with any of the other services. What was meant to be a cost savings interoperable system, was another waste of government funds. The system could only function within the confines of the Air Force domain.

Because of this simplified, interoperable system we were unable to pass the draft MOA to all the services at once. I had to revert to the "manual" system of playing an email game of "Telephone." After I consolidated comments, we had an original "host" document. I had to send the host to one service who had to internally staff the document. Once that came back, then email it to the next, and then to the next. I can't remember the suspense date, but for the most part extensions were required. And each service took their time in staffing the document. We kept pushing the MOA to the right. I had to ensure that for the document to be legally acceptable, we needed an original signature from each service general/flag officer —not a copy. I needed one true document. Round and round we went, passing the document back and forth, over the course of six months.

All the while, my leadership was breathing down my neck to get the document finalized. I still shudder when I hear those three useless letters. A PhD in Astro physics must have devised TMT. It was neither user friendly, nor easily understandable. But in my office, the determination of my success or failure was not based on my operational, tactical, or strategic capabilities. It was based on how well I could navigate TMT. Supporting war fighters and establishing sound policies were secondary. If you didn't know TMT, your performance rating was already teetering on substandard. And for me it was. As I approached the end of my career, I was informed that because of my failure of AF staffing, which was mainly my total ignorance of TMT that I was going to be placed on a Performance Enhancement Plan (PEP), meaning my performance was sub-standard. Of course, by people who had no experience in actual global MWD operations. This was performance optics, not performance evaluation. Leadership capitalizing on my weaknesses not on my strengths.

The MOA itself was useless…and incoherent at best. It did not serve or meet the needs of the services. It was just another method of Air Force control. The voices of the JSMWDC team, as always, were absorbed in the halls of the fifth floor of the Pentagon. The EA failed to listen to the service recommendations, mainly due to the layer of oversight that was the brick wall between me and the EA, unless it was presented by the Blue Team.

Chapter 5 – True Colors

As the DoD MWD Program manager, I ran Joint Staff issues to the Lt. Col. who was my branch chief. For the most part, any idea that didn't fit the "ghost guidance" of the EA's core tasks was curtailed on the spot or elevated to the Division Chief, an Air Force Colonel who openly stated his lack of interest in the MWD program. It would rattle around his office and then be returned as unimportant to the EA mission.

During the period that I had been recalled to active duty from January 2005 to October 2012, I had the honor and privilege…along with the greatest MWD handlers, trainers, and Joint Services Program Managers…of building the U.S. CENTCOM MWD Program. For just shy of eight years, I was the catalyst, executing a program that provided over fifteen hundred military, contract, and coalition dogs to the war fight. Initially, we had no policy, no deployment tracking, and no established canine chain of command. Along with my canine mentors—all enlisted non-commissioned officers (NCOs)—we framed the program. Together, we resolved multiple issues—policy development, pre-deployment training, establishing transit hubs (McGuire & Kuwait), utilization reporting and accountability. We lost thirty-two MWDs and five handlers during my tenure, and to this date, I remain the only person to have executed an operational canine program in a theater of war.

Yet, my Branch and Division Chief—with no canine experience or strategic direction for dogs in a combat theater—overrode experience, knowledge, and strategic vision. Neither the EA—nor any previous EA—had served in a Joint Assignment. They all served under the protective umbrella of the Air Force. I lived in a "purple world" (Joint) for almost eight years. My Air Force leadership lived and breathed one color "Air Force Blue." However, the world changed over the last century. The U.S. has never gone to war independently since WWI. Our military worked in a different reality…Joint.

Unfortunately, no good deed goes unpunished. Shortly after what would be my final JSMWDC meeting in October 2019, I underwent surgery for an aortic aneurysm. It was discovered in 2016 and I'd been under a cardiologist's care. In August 2019, during my semi-annual echocardiogram, it was discovered that my aneurysm had grown to the point where the cardiologist recommended surgery. I hit a home run. I had never had surgery, minus tonsil removal, when I was six. I was nervous and very anxious. The verdict after a surgical consult was that the procedure to correct the aneurysm needed to be performed no later than the end of the year.

We scheduled the surgery for November 19th and fortunately the procedure went well. I was released from the hospital on November 23rd and went home. To my chagrin—and not surprising— neither my Branch Chief nor my Division Chief, my direct supervisory chain, contacted the hospital or my wife to check on the outcome of my surgery. Not one call. Leadership. I'll use the word…disgraceful…and refrain from my New York vocabulary describing what I thought—a swath of profanity hopefully made their ears ring. The EA did not contact me either. However, our Deputy, a wonderful visionary and leader, reached out. My guardian angel in many respects intervening in 2018 when after seven years of receiving an annual performance rating of outstanding, my division chief assessed my performance as "met standards." Without the Deputy leading the way, the DoD MWD Program would still be in the "stone age." The other, a lady who I consider a very dear friend. Ms. Bridgett Simpkiss, or as she is known to all as "Gigs," frequently reached out to check on me with kindness and concern. To both of you, a sincere thank you. Your kindness to me and my family helped in my quick recovery.

I was ordered home for an eight-week convalescence. I received no calls, no cards, no fruit basket from anyone in the division. No one cared. I could have died, and it wouldn't have mattered. I did not accept the Air Force culture. I didn't speak AF. I didn't believe in their single-focused mentality and inability to think strategically. It was easy to see the message. I did not play the Air Force game. However, I was extremely blessed in that my Army family circled the wagons, calling my wife daily, sending me texts and well wishes. Even Major General Glaser, the former Provost Marshal General of the Army, called to check on me, "Anything I can do? Call me." His sincerity was true, and my admiration for the General and my former Army team grew even stronger.

Where was the Director of Security Forces (EA)? Certainly not caring about one of his troops. When I got off the phone with General Glaser, I wept. I was so thankful that he reached out—and depressed that the Air Force had no concern for my welfare. My former supervisor and very dear friend, Tom Blair even came to visit me in the hospital—and maintained a constant vigil on my recovery.

I'd established and executed a combat canine program in twenty-six countries for close to eight years. But I did not wear the Air Force blue and had called out the shortfalls and narrow-mindedness of Air Force EA oversight. Years of experience and operational engagement were unimportant in ensuring the ability to support global missions and protect our troops.

Remember I was a branch of one, and the mission needed to continue. By the time I had gone for surgery, a contractor was hired, but was still very new to the MWD program. From the day I got home from the hospital, I was already starting to look at emails, and get back into the game.

In January 2020, I was still convalescing. Still engaging, but after a heated email to my branch chief, I received a reply that crushed me but didn't surprise me. For all that I had done over the course of the span of more than fifteen years…being in the thick of canine operational, tactical, and strategic missions… all became null and void. His email informed me that I had gone "native," and that I was working outside of my authority. I was not accomplishing the "core" responsibilities of the General (EA). Subsequently, there was a follow-on email asking me about my plans for returning to work or was I planning on retiring? If I was going to return to work, I would be placed on a performance enhancement plan (PEP). A PEP is for substandard performers. Meaning the first step to do one of two things, first establish the benchmarks for having me terminated, or second to reel me in and have me drink the Air Force cool aid. This was an Air Force preemptive strike. The handwriting was on the wall. My rogue behavior created too many issues that the Air Force didn't want to address.

This all aligned to a behind-the-scenes initiative to upgrade the position of DoD PM from a GS 14 to a GS 15. This was something that I didn't know at the time. Regardless, in good conscience, I couldn't return to an organization that had no concern for my well-being or held me accountable for failing to follow "Core" values. The issues I had …. what were the CORE Values? No one ever presented them to me. And when I provided the General with my After-Action Report in April 2020, his response in a detailed email was that he was unaware of half the issues that I was bringing to his attention.

I decided to retire. While I was making that decision, there was an effort by two very distinguished gentlemen to intervene on my behalf. Mr. Dave Kontny was the Chief of Staff Joint Program Office for Countering IED for the FBI and Mr. Radoslaw Olzewski Director of the EU Aviation Counter-Explosive Division. Dave has been my canine mentor for more than twenty-five years. He instilled in me the importance of inter-agency collaborations and going outside the box. "Rad" was a godsend. He pulled a rabbit out of a hat when we were unable to get our deploying U.S. Army Europe canine teams to the Middle East, imprinted on IED components (TATP and HMTD). Through Dave, Rad arranged for the Belgian Federal Police to imprint our teams. He also opened doors for our Army canine teams stationed in Europe to train with the Portuguese Federal Police and attend European Union counter-explosive meetings. Those initiatives were considered not part of my "ghost" CORE mission.

We opened many doors in partnership and friendship. Dave and Rad both asked me to reconsider. Dave said he would intervene on my behalf. I didn't want to let them down and was hopeful that there could be some "negotiations" allowing for a more palatable work environment. Dave set up a meeting the week that I was scheduled to return from convalescent leave. To be honest, I was hoping that we could set the record straight and move forward. But not surprisingly, and to my chagrin, the Air Force Lt. Col. cancelled the day before we were supposed to meet. No reason. No rescheduling. Just cancelled. That solidified my decision to retire, and nothing was going to change my mind. That same Lt.Col., has since been promoted, using the title of DoD MWD Program Manager, for which he had absolutely no knowledge to add to his promotion credentials. Yet there was no collusion or pre-selection in *that* process.

Powell-ism

"Leadership is the Art of accomplishing more than the science of management says is possible."

No one in history has defined leadership as well as General and former Secretary of State Colin Powell. His candor and leadership profile has been a forerunner of developing successful organizations within the government and civilian sector. He knew that redundancy in an organization would lead to failure. In the DoD canine program, General Powell would have been labeled as I was, as having gone "Native."

Powell embraced the need for evaluating the future. Change was endemic to an evolving organization. Accepting change, well that's another story. He commented, "I'll bet you right now that there's no established organization where you won't find somebody who says…I know what I've been doing for the last fifteen years, and you're not going to screw me up."

That was the true essence of the DoD MWD program. I had led a canine mission in a shooting war, and I'd learned many lessons from my colleagues. More importantly, as a collaborative team, we had established a strategy using canines for counter IED mission maneuver units and easily adapted by first responder canine teams.

I learned that as a good leader, you must make decisions. That's what you get paid to do. That's why you're in that position. If you can't make decisions, you have no business being a leader. Go somewhere and stick your head in the sand. And I pissed people off. I didn't sugar coat facts that needed to be presented to senior leaders so they could make sound decisions. To get the job done, while—more importantly—protecting our young men and women. Many times, I made decisions based on situational awareness, and information from great partners who were forward deployed. My job was to support them. I led from the front. When I made a decision, if it was verbal, I also informed that person to make sure that they recorded the date, time and specifically who was giving them that order. I stood by my decisions. In my office, there was a huge disconnect between bureaucratic management and getting the job done leadership.

The War Years

Chapter 6 – Operations Iraqi and Enduring Freedom

I was blessed and humbled to serve with the greatest members of the Armed forces during the time I was recalled to active duty (2005 -2012). They were dedicated, honest and committed to the mission and their brothers and sisters. Not a day goes by that I don't reflect on our great team in the CENTCOM MWD program.

We focused on one mission—keep our troops safe! Probably because it was wartime, the redundant bureaucratic games, death by PowerPoint and the atrocity of TMT that are played in the Pentagon, didn't exist. I was expected to make decisions and execute the mission. I had the total support of the CENTCOM Senior Leaders. I earned their trust under the most difficult of circumstances. As a result, this great canine team set in motion an MWD strategic and operational template (NATO STANAG 2623), a foundation that could easily be adapted and tailored to meet any contingency around the globe.

Overview

Before I go into the CENTCOM MWD story, I wish to acknowledge the Theater and Service Program Managers, Kennel Masters, Force Managers, Trainers, and Handlers. It was total unity of effort from the ground up. A team effort! We divided and conquered. I provided the field with policy guidance and ran interference with the multiple agencies needed to administer the program, while they executed their missions on the ground. They were my mentors. And, to my gratitude when I went off track, they "gently" refocused my energies. They made me proud, and I was honored to serve in this position for almost eight years. They are my heroes.

CENTCOM was comprised of the joint military services and a 26-country partner coalition. Every individual in CENTCOM contributed to the War fight. When you arrived, you weren't given a chance to breathe. You set your priorities on the run and started executing the mission immediately. You were expected to adapt and overcome. The Operational Tempo (OPTEMPO) didn't stop. Twenty-four seven, the mission forged ahead like a freight train. The war didn't stop. You learned quickly. The lives of soldiers, sailors, marines, airman, civilians and contractors depended on your decisions. One thing was for sure, if you couldn't keep up with the Operational Tempo (OPTEMPO), your time at CENTCOM was limited.

I arrived at CENTCOM in January 2006. At that point, the priority was to train Iraqi Security Forces and rebuild Iraq's infrastructure. Having just returned from Commanding the Iraqi National Police Academy, I was aware of Muqtada al-Sadr's edict. As leader of the insurgents, he had issued a jihad against coalition forces. The mission transitioned from expelling Saddam Hussein to reconstruction and fighting an insurgency. To meet this tactical demand, the U.S. deployed more and more forces and equipment. And the theater grew exponentially.

The public and national command authority pressured the military to make progress in Iraq. Counterinsurgency doctrine was written as we attempted to rebuild Iraq's security forces. It was a unified effort, with the State Department trying to develop a democratic government through the Iraqi Reconstruction Act.

Dogs quickly became high demand commodities for maneuver commanders, facility force protection and contractors operating at remote sites. While there was an increase in demand, our military and civilian leaders failed to recognize that dogs were not energizer bunnies. They required the same care and rest as our troops to be effective in combat. The need far outweighed the inventory.

The Beginnings

When I reported, I was initially assigned to the Joint Security Office Detainee Operations Branch. That was a one hundred percent reversal from the mission I'd just performed. I've never been a good staff officer. Adapting to formats, fonts, power point and staffing actions were difficult for this old soldier. My computer skills remained at the basic level. I'd rather go face to face and present my response, than waste hours upon hours putting together the staff packages and legal opines necessary to complete just one action. I was overwhelmed and frustrated.

Fortunately, the CENTCOM Provost Marshal…a methodical West Point full Colonel, who led from the front and exuded the importance of setting the example…realized that this was not the place where I could "make a difference." As fate would have it, one of the young MP Captains resigned his commission to accept a position as a special agent with the FBI. Rather than continuing to allow me to foul up this highly visible mission, my division chief had mercy on me. He gave me the opportunity to assume the Divisions' MP force management responsibilities. I welcomed this transition. I had previous experience in the force management arena when I was assigned to the National Guard Bureau in the late 1980s. The change propelled me from the frying pan into the fire—a fire that burned 24/7, 365 days a year. I was totally unaware of how critical this duty was.

I became the MP unit doctrinal—movement, capability, pre-deployment, and joint service - coalition subject matter expert. This was not only for military police assets, but for the convoy security forces units, i. e. forces created from artillery, transportation, air defense, and other units whose mission sets were not in high demand. The mission was to protect military and civilian logistics convoys. It was dangerous, but bottom line, we modified and adjusted to meet the mission. In this role, I was responsible for matching requests with capabilities. I screened every request for forces from Iraq and Afghanistan. We did the deep dive and looked at the MWD inventory to see which service had the capability. Same with detainee operations, and police training. The MP enterprise was the most "Purple," in that every service contributed to the holistic MP mission. This was one of the most rewarding chapters in my life. I was at the tip of the spear, part of the hammer to execute CENTCOMs strategic goals and objectives.

Military units deployed on a rotational basis. For most units, the deployment cycle was twelve months, with the Air Force deploying in increments of six months. The number one mission was to ensure that all units arrived on their scheduled LAD (Latest Arrival Date). A unit unable to meet their LAD—for whatever reason—threw a monkey wrench into the movement system. It also reduced the actual number of days the incoming unit had with the outgoing unit to accomplish their transfer of authority, or their left seat right seat transition. It was a cardinal sin to miss LAD. However, there was no playbook or operational guidance for the deployment of dogs. And there was no guidance for in-transit support, no theater MWD policy or utilization guidance.

That was a tremendous task that needed to be fixed. Thankfully, with a team of "real" canine professionals…all NCOs…to guide "Pops," we resurrected and up righted the program that had become a self-licking (no pun intended) wound.

The main elements that were deploying consisted mainly of Army Brigade Combat Teams (BCTs) and the Marine Corps Marine Expeditionary Forces (MEFs). They performed outside-the-wire counter-insurgency operations. They faced threats from deadly improvised explosive devices (IEDs) planted as roadside bombs, causing numerous casualties to both vehicle and dismounted patrols. The only capability known to detect and deter these heinous explosives… the military working dog!

Dog teams, a handler and dog…like our major units, normally a battalion (600 – 1000) personnel, and Brigades (4000+) …were considered a unit. These MWD teams were tracked just like the large troop units. Like a social security number, each unit is assigned a force tracking number, commonly referred to as an FTN. This number identified the team's deployment cycle from start to finish. Whenever someone wanted to look up your unit and obtain deployment or mission information, they would type in the FTN, and that information would appear on the screen.

While large-sized units had a chain of command with seasoned leaders, the dog team was no more than a young man or women, usually between nineteen – twenty-three, a lower ranked enlisted troop and his/her dog. They were on their own, having to coordinate transportation from their home kennel in the U.S. to the port of debarkation. These young troops were responsible for the health, safety, and welfare of their MWD. That included moving approximately 500 pounds of canine gear (Vari Kennel, five days of dog food, medicines, bowls, leashes, and first aid kit) on their own. They also moved their individual soldier gear (weapon and load bearing items including helmet and fifty-pound body armor) plus any other special equipment for their dog. Many times, they were bumped by a 300+ unit.

There was no canine policy or playbook. What was needed was a directive that would ensure smooth cyclic rotations. First, commanders needed to be educated about what these dog teams could do to support their mission. The canine chain of command and their logistics support needed to be clearly communicated. And, of course, MWD capabilities, utilization and reporting requirements all needed to be spelled out. Our MWDs were all attached primarily to infantry units. Everything they needed had to be provided by the infantry unit they were attached to. Dogs teams had no vehicles, no communication equipment, no counter IED equipment. And that was another problematic situation. My flow of information and canine intelligence was through my daily contact with the Military Working Dog Program Managers in Iraq and Afghanistan. They were the finest Non-Commissioned Officers I've had the honor to serve with. As I continued to reach out to our deployed Program Managers, I also engaged the Army, Navy, Air Force and Marine Corps program managers quickly gaining an education on their service MWDs, capabilities, training and learning there force management process. The critical shortfall… the need to physically track and provide in transit support during movements overseas. This requirement led to the creation of the "MWD Transit Hubs." Frustrated, one day I was replying to an email from one of the program managers. As I was concluding the email…and it seemed that MWD has fallen under my area of responsibility… I added in my email signature block:

Lieutenant Colonel, MP
CENTCOM MWD Program Manager

From that date sometime in 2006, I assumed ownership of the program. That provided a clearing house for information and positional authority to act on recommendations from the field. What was needed first was direction. Our first task was to roll up our sleeves and start devising a canine directive. Retired Sergeant Major Jeremy Peek was the founding father of the CENTCOM MWD Policy. He and Sergeant First Class Rene Johnson the MWD Program Managers in Iraq worked tirelessly in framing the guidance. Jeremy…who has been my mentor for years…was a policy wiz. But as the Combat Theater subject matter expert, he and Rene ensured that the draft policy was founded in the best MWD practices to support our teams in transit and in the field. The draft was passed to the other MWD stakeholders. It captured MWD deployments, including the mandatory requirement for teams to attend pre-deployment training It also elevated the blood pressure of deployed commanders. The policy placed ownership of all service MWDs under the MWD Theater Program managers. Finally defining an MWD Chain of Command.

For close to a month, we worked on this. It was a document built by NCOs. However, to get it published, I had to manipulate the system a tad. First, I inquired as to the procedures to get this published and ran into brick walls. Not intentionally, it was just that CENTCOM had directives and regulations, but no policies that assumed oversight of an operational mission. That was unique. However, if you try hard enough, you can always find a solution, not the perfect one perhaps, but a solution that achieved the seal of approval as a CENTCOM official document. "Administrative Guidance." It was a clarification of an already existing operational mission. It didn't have to be staffed. And although it technically provided policy guidance, I was able to skirt the staffing process. Not intentionally, but I achieved the purpose of the document…to fill the void and provide CENTCOM MWD guidance to our Commanders overseas.

I had pushed the envelope. No one said no, and I took that as drive on…and got the mission done. It was a living breathing document. If changes needed to be made, it could be modified immediately and published in the Automated Military message system.

We continued to modify and refine the "guidance" to meet the needs of our teams. Case in point, when SFC Peek rotated, SFC Rob Sanders came in as his replacement. I'd tasked Rob and his constant companion SFC Rene Johnson (we were lucky that Rene was not scheduled to rotate for a few months), to develop a reporting template that tracked the operational status of ALL dogs deployed in CENTCOM including U.S., coalition and the myriad of God knows how many contract working dogs.

Rob and Rene's reporting format captured the full spectrum of our dog teams that were disbursed throughout CENTCOM. The report provided leaders with answers to tactical questions: How effective were our dogs in detecting explosives, what areas in the country were most of the finds, how many teams were injured and by what? This template would be the tracking mechanism that would finally provide contract working dog oversight, ensuring that all dogs in our backyard came under our command authority. This report provided strategic analysis and accountability. The MWD Program Managers owned the dogs. They knew the numbers by command/service, types of missions and hours utilized. It was critical in determining canine effectiveness across the Area of Responsibility—and another step in unifying full spectrum theater canine operations.

Rob and Rene were the Picassos of this masterpiece. The report was shared with our Commanders, the Military Services, DoD Executive Agent, and the Joint Staff. It was the only report identifying canine utilization and effectiveness in CENTCOM.

Soldiering On

And the mission kept on rolling along. I had force management responsibility for approximately 1/3 of the units assigned to U.S. CENTCOM. Hard to believe, but true. While we had up to 14 Infantry Brigade Combat Teams and Marine Expeditionary Forces units, each having a force tracking number (FTN). Each two pax dog team had an FTN. I had to track 600 dog team FTN's, as separate units. Each one needed to be verified and accounted for. Not to mention the MWD's we had providing Air Base Defense and force protection at our bases in Kuwait, Qatar, Bahrain, and in the "Stans."

This labor of love was not without significant challenges. We needed to work with the services to determine a basis of allocation. How many dogs were going to be allocated to each unit? We realized that the demand was increasingly placing a strain on the supply.

Add to this, the Department of Defense deployment cycle. It was a one to three ratio. For MWDs as well as combat forces, it consisted of deployment of one third of the forces, one third in training to deploy, and one third would be in what we called "Dwell,"—a year of rest and recuperation. And that period of Dwell was UNTOUCHABLE.

Strategically, was the supply chain of DoD produced dogs meeting the increased demand? No. While requests for dog teams poured in, we were operating on a shoestring inventory. The DoD Executive

Agent had not increased production to meet the growing need and shortage of MWDs. They had not planned how to replace wounded and dogs killed in action.

This created two major challenges. First, the DoD MWD School operated in a pre-war production mode, a stagnant 1999 methodology. Even today, the EA has nothing in place to expand MWD production to support surge requirements. The EA demonstrated no interest in the performance and effectiveness of dogs operating in Iraq and Afghanistan. Procure and train dogs for the services was…and remains…their mantra. However, this ignores tactical relevance, situational awareness, and strategic vision. That lack of vision and void in strategic thinking led to the most disgraceful chapters in the history of the MWD program. Stay tuned!

We had to look at the "pool of assets", the priority of effort providing explosive detection dogs to the theater. We had to look at canine capabilities not the color or national orgin. Working with the Joint Staff and the service force providers we commenced utilizing all available assets to fill the needs of our deployed combat units. Air Force and Navy dogs would plug the canine gap, trading in the Wild Blue Yonder and the Deep Blue Sea, for Army Desert Camouflage. There was no other choice. That decision…saved countless lives!

The next step was how to integrate non-Army dog teams, train them on Army canine tactics, techniques, and procedures or TTP's. Refocusing team mission detection mind sets from Navy/Air Force MWD roles of installation force protection to being the tip of the spear. Leading Army patrols outside the wire, in the direct line of fire while searching for IEDs in 130-degree weather. This was even more complicated as these non-Army teams would be thrust into the "individual augmentee" deployment system, removed from their services protections, moving from their home station to the middle east without a canine support mechanism. We needed the "HUB."

The Marines Have Landed – The Situation is Well in Hand

The pre-deployment training of Air Force and Navy teams deploying to support Army combat units, came in a Christmas package from the Marine Corps. The Marine Corps Warfighting Laboratory and Combat Development Command were visionaries, and along with Mr. Bill Childress and Mike Wells, the USMC Canine Program Management Team, saw the need to develop a training site where Marine MWD teams could be introduced to the rigors of combat. Training in a scenario-based environment mirroring the conditions that they would face in Iraq and Afghanistan. Yuma Proving Grounds (YPG) in Arizona was where the Marine Corps established their MWD canine pre-deployment training site. They had invested close to a million dollars to create an Iraqi and later an Afghan village. YPG, as it was commonly known mirrored the deployment environment, with temperatures climbing up to 130 degrees.

Bill and Mike along with Gunnery Sergeant Massey played no games. Gunny Massey on day one had handlers gear up in their combat gear, with their MWDs go on a five-mile conditioning run. Welcome to YPG! The Marines were preparing their MWD teams right from the "get go." We needed to get on the bandwagon and make this the premier MWD Pre-Deployment Training site for every dog team that was going forward. And we did.

These "kids" would be able to hone their team skills, which included mission planning. That was extremely important. Our handlers were lower enlisted ranks. When deployed it was their responsibility to present a mission brief to senior Non-Commissioned Officers, Junior, and Senior Officers. The handler was responsible for their dog. Leaders needed to be aware that the Handler, not the unit leader made the decisions for their dog. Most of the leaders needed to be educated to understand the capabilities and utilization of the MWD team. The Marines saw the need for all MWD teams to be trained and conditioned to the deployment area of responsibility. They stepped up to the plate and opened YPG to the services MWD teams. The services bill payer, assign a liaison NCO providing Army, Navy and Air Force admin support and command and control.

The expansion of YPG, directly supporting the war effort, would be funded through SECDEF Overseas Contingency Operations funding. (Justification: being in direct support of a designated named contingency operation (OIF/OEF).) The YPG pre-deployment site remained in operation until 2014. By that time, MWD operations overseas had dramatically been curtailed, and the SECDEF contingency funding had stopped. Operating costs returned to the Marine Corps. YPG demonstrated the importance for MWD pre-deployment training, and it established the template for pre-deployment training for future contingencies. God help us if we fail to provide MWD pre-deployment training. If we turn a blind eye and put them into theater ill prepared to face the threats and unforgiving environments, we face the unnecessary loss of irreplaceable force multipliers.

Capitalizing on "Administrative Guidance," we modified the document establishing YPG as the CENTCOM MWD Pre-Deployment Site. MWD deployment orders mandated that ALL MWDs attend pre-deployment training at YPG. Initially, the Marine Corps assumed scheduling, training, and command and control for all the services. But as the service liaison NCOs filtered in they picked up the task of scheduling their dog teams. To help Santa, we also had Santa's helper…the Bureau of Alcohol, Tobacco and Firearms. Bill had the ATF voluntarily come to YPG and introduce our dogs to the terrorist explosives of choice, Triacetone triperoxide or (TATP) and Hexamethylene Triperoxide Diamine (HMTD). The components were easily attainable and extremely volatile. Even today in our most recent attacks in European Countries (France, Germany, Belgium), TATP and HMTD were the explosive of choice. The components are available from commercial sources.

The ATF would come down for a week with their chemist and ATF canine trainer, to introduce our military working dogs to TATP and HMTD. They provided the platform for the dog to be "imprinted." The process where the dog's olfactory memory retains the scent, what we call "muscle memory". Thankfully, the ATF was instrumental in adding this enhanced lifesaving capability. The CENTCOM canine mission was accomplished due to the Marine Corps opening its doors to our joint service dogs. We owed the ATF a huge debt of gratitude. They joined the war fight as brothers in arms, preparing countless dogs to keep troops alive. At the height of the war, close to 600 joint service teams were rotating through YPG pre-deployment training annually.

Aye aye, Santa! Semper Fi! You trained these MWD teams on the threats and situations they'd see during deployment. A part of canine history that saved hundreds of lives…hopefully not lost to the ages!

Pre-Deployment Inter-Agency Collaboration

Sandia Labs – A Missed Opportunity

While serving with the Office of Provost Marshal General, we developed a solid foundation with our counterparts in the Explosive Ordinance community. Working side by side, we wrote the revision of the Joint Staff Counter-IED Joint Publication 3-15.1. Counter Improvised Explosive Device Operations [Reference 14] published it in January 2012. This was the first time that MWD utilization, capabilities, and procedures were included in this joint publication.

This led to a unique opportunity. The MWD community was engaged with the ATF and FBI, supporting home-made explosive training. Bringing explosives onto army installations required clearances from the installation commander. The Army's EOD Policy Board determined these requirements. To my amazement, the Army EOD policy board worked collaboratively with Department of Energy (DOE) stakeholders at Sandia Labs New Mexico.

The DOE provided studies on the effects of Triacetone Triperoxide or TATP. TATP remains the most widely used home-made explosive in the world. The Army also used Sandia Labs for technical pre-deployment training to introduce deploying EOD technicians to the scientific configuration and volatility of TATP and other explosive compounds.

This was an eye opener for me. EOD technicians received the full flavor of the volatility of TATP as they all had to perform the "hammer test." A minuscule amount of TATP—approximately the size of a pinch of salt—was placed on a metal dish. When the technician hit the compound, the volatility of the explosive got your attention. It was equivalent to what we knew as kids as a 4th of July "cherry bomb." The lab was open to allowing Army MWD handlers attend the same course, providing them a better scientific understanding of IED's, and the fun of performing the "hammer test."

This meeting was the foundation that led to the Army EOD Policy Board amending the transportation directives for home-made explosives and components on Army installations. We had another platform for integration of training and extended explosive research and development. We again opened the interagency door, unfortunately an opportunity missed due to lack of funding, Sandia Lab was just a bridge too far.

Force Flow – Transportation

How do we smoothly plan and move MWD teams to and from overseas? At the beginning of Operations Iraqi and Enduring Freedom, handlers were on their own to make their way from home station to Dover Air Force Base, the cross-channel movement port of embarkation. The Air Force provided flights for units (battalion sized) to the central receiving facility in Kuwait. These troops were processed by Army Central Command (ARCENT) where they would complete acclimation training for two weeks prior to onward movement to Iraq and Afghanistan.

In many instances dog teams arrived at Dover only to be bumped, as they did not fit the profile of a typical unit. A young trooper with his/her dog…according to DoD…was a separate unit. The best many of the dog teams could hope for was to be manifested as part of a standard organization and "hitch a ride." Adding to the confusion and need for centralized canine force management. There were no temporary kennels, or temporary lodging. Handlers had to pay out of pocket for off post lodging for as long as it took to get a "seat" on a cross channel flight. Because we had no formal MWD liaison, dog teams had to wait…in some cases ten to fifteen days to get a flight.

There was no MWD deployment guidance and no procedures to attain veterinary support. They were in "canine limbo." Add to that that MWD teams are expected to train, to keep their dogs "on odor." The Dover kennels did their best to help, but there were no formal agreements, it was all ad hoc. There just wasn't a coordinated plan. The longer the dog was unable to train on odor, the possibility that the dog would lose olfactory capability, effecting the ability of the dog to work. The other concern was that, by regulations, if a dog does not train for a 30-day period, it is required to be recertified.

Army veterinarians assigned to Dover AFB were only responsible for providing medical care for the assigned Air Force Security Forces MWD and dependent pets. Veterinarians were not staffed or resourced to provide Veterinary support for the influx of deploying MWDs. MWD teams required ten-day health certificates to deploy overseas, and normally be provided a cursory physical examination. Without dedicated veterinarians supporting deployments, the MWD health was in jeopardy. In some cases, we did have dogs arrive to their supporting units in Iraq and Afghanistan that were medically unfit to perform their mission, placing troops in jeopardy of becoming casualties from improvised explosive devices.

To say the least, this was a dilemma that no one thought of. A handler and a dog were not on anyone's radar screen. To make matters worse, once dog teams arrived in Kuwait, no one knew what to do with them. Once "Units" arrived in Kuwait, they would automatically be sent for two more weeks of acclimation training, prior to onward movement into the combat zone. Dog teams didn't need additional training, they were certified prior to deployment at home station. However, as there was no canine reception plan, the teams were packed up and sent with units for acclimation training.

These teams—having dealt with the agony of independently moving themselves from home station to Dover to Kuwait—now faced more delays. MWD teams travel with five days of dog food and medicines. They are totally reliant on the veterinarian and kennels for resupply. That wasn't happening. MWD teams shipped to "acclimation facilities" were on their own. In some instances, our MWD teams were "lost" for up to forty-five days. Something had to be done and that something was my responsibility. But how would we develop the mechanism to take care of our MWD teams during this arduous process? We knew what we had to do as leaders…take care of our MWD Teams!

The Icing on the Cake

You can't lead from behind a computer in a cubicle. I made it a point to get into the theater as much as I could to meet the team that I was supporting. The troops on the ground were fighting the war. Equating my job to a movie, I was the producer and director, operating behind the camera. But where I had the ability to influence the operational mission was having the positional authority, representing the Four-Star Command. I could guide and influence the strategic enterprise of MWD operations with senior leaders, General and Flag Officers.

While I was in Qatar attending a Force Management conference in 2007, I was able to coordinate a quick trip to Iraq to get a heads up on the MWD operations from SFC Peek and SFC Johnson. It was an eye-opening experience. I needed to see firsthand and…most importantly…learn what was happening on the ground to address the transportation problem.

We had a phenomenal chain of command, but the most influential member of the chain was that young MWD handler. They walked the canine walk and talked the canine talk. Their lived experiences were where the rubber met the road, and if I were going to be the best advocate for the program, I needed to follow the flow firsthand. As General Powell stated, "Truth pays dividends." The truth was about to blow me away.

Meeting SFC Peek and Johnson, the "Authors" of the CENTCOM best seller, was a privilege. I should have asked them for their autographs. Now, it was time to get my hands dirty and talk to the handlers, what they were experiencing, what more did I need to do as a leader to support them.

Fortunately, there were three handlers who had just completed their tours and were getting ready to rotate back to the states. SFC Peek was kind enough to bring me over to the transition area where he introduced me to two female Navy handlers, and a young Army handler who decided he was going to be anti-social. I got a good briefing from the Navy handlers. Their concern was that no one was tracking their arrivals, and the supply chain for MWDs was non-existent. The Navy handlers were PEDDs (Patrol Explosive Detection Dog Handlers) and were pressed into service in the Army, serving with an Army BCT. They were thrilled to have performed a "real" mission and both teams had several "finds." The Army loved our sea-going handlers and took exceptionally good care of them. Fortunately, these two handlers deployed together and were returning together as "battle buddies."

Then there was my "problem child." I am not a person who hides my emotions. I'm from New York City, and we New Yorkers are proud to be rather obnoxious and abrasive. I've never known why. We just say what's on our minds and if you don't like it you can go "F—K Yourself!" To my surprise, this "unique" quality followed me throughout my life. For some reason, I must have made a difference. For years after I left active duty, I would inadvertently run into someone in the Post Exchange, or commissary telling me, "Sir, I remember you, you used to tell people who had no clue about MWDs to "Go 'F' themselves." Even in retirement, my Mantra remains the same. I've told my wife that when I pass to make sure that this is indelibly inscribed on my headstone.

Setting the stage for my Army handler "chat," SFC Peek and I entered the tent… he introduced me to this young Army specialist. He didn't stand or acknowledge my presence. He just mumbled under his breath, which made me angry. I addressed him and got the same indecipherable, muttered response. I then asked SFC Peek to leave the tent. The next several minutes, I filled the air with inexcusable profanity and vile language. I made my point crystal clear, that number one, I was probably old enough or probably older than his father (I was fifty-one at the time) and two, that I would beat the living s**t out of him if he didn't talk to me. Political correctness from a cop was never my strong suit.

At that point, he finally decided we would have a chat. I started with, "What's your fucking problem?" I had calmed down a tad. But I was about to be blasted with the reason for his attitude…and this young soldier was about to break my heart with his story.

This young man was a new handler on his first deployment, excited, motivated, full of piss and vinegar. He wanted to get to Iraq and be the tip of the explosive detection spear. Then he got caught in the transportation maze. He flew from home station to Dover where no one picked him up in Philadelphia. He paid out of pocket for the cab fare to get him, his dog and 500lbs of equipment to Dover. Re-stating, there was no canine reception at Dover…no one was tracking his arrival, there were no kennels or temporary lodging. He had to pay out of pocket for a cab daily to get him back and forth to his off-base hotel. And the hits just kept coming. This nonsense played out every day. He had to go back and forth to the Air Terminal to find out if he was manifested on a cross channel flight.

Then, when he finally got a flight to Kuwait, no one had a clue what to do with the team once they arrived in Kuwait. Because we had no process in place. He was shipped off to one of the acclimation sites. They should have been put on the first deploying aircraft to Iraq.

I looked at this kid and listened. He was my responsibility just like my own kids. All the handlers were my children. His only job was to provide life support for his dog and get transportation to his unit. He had a five-day sustainment package (dog food – medicines) with him. Because there was no canine tracking mechanism in place, this critical lifesaving asset was delayed in Kuwait for eighteen days. When he was finally able to get air transportation to Iraq, it was not through the Army, but by getting assistance from the AF kennels at the adjacent Ali Al Salem U.S. Air Base. Finally arriving at BIAP (Baghdad International Airport), SFC Peek took charge and got the team to their unit. A horror show.

At the time, he needed a full resupply of preventative medicines, science diet (canine dog food), a new leash, and a new vari kennel. He needed a canine overhaul, and he hadn't even started his mission.

I sat there motionless, asking stupid questions. Who was supposed to help you? Couldn't anyone track your deployment number (FTN) and avoid this unnecessary delay? What was I thinking? There were two systems, the Air Force movement system, responsible for getting troops forward, and in this case, Army Central Command, whose mission was to move "units." Process incoming units and get them to acclimation training and into the combat zone. For dog teams no one had included a processing plan, because at the time, there was no MWD leader directing the process. And there is no fault attached. It was an oversight, an oversight that was about to be corrected. But the question that I pose to you today are dogs on anyone's planning radar? Someone needs to ask that question.

After letting me have it with both barrels, he yelled back at me. "Sir, I know I'm going to get fucked going home, no one is going to take care of us. We're going to be treated like shit and it will probably take us thirty days to get home!"

This hard charger was not worried about the year he faced being on point in front of troops looking for IEDs, possibly coming under direct enemy fire, or placing his life in harm's way. He was angry, disgusted, and frustrated because there was no MWD chain of command that was going to take care of him and his dog. All he was looking at was another long-drawn-out nightmare of "getting fucked."

The chain had failed this kid. Under my positional authority as the CENTCOM MWD Program Manager, his problem was not his anymore. It was mine. And it was going to get fixed! The problem was and remains that after every war, dogs become low hanging fruit. Templates for program management get shelved and forgotten. We did it after WWII, Korea, and the debacle of leaving dogs in Viet Nam. We were…and still are…not prepared. You'll see that with all of the concerns over the past few years about possible action in the Korean peninsula, INDO-PACOM has zero plans for the utilization, employment, movement or resupply of dogs. Should the shit hit the fan…and it will somewhere in the world…the military will trip over its shoes trying to figure this out, all over again, starting from scratch. The lessons learned from previous conflicts will have been all but forgotten.

The truth of the matter, besides the handlers, veterinarians and those directly engaged in the operational canine mission, the MWD program is not even an afterthought in the minds of the Joint Staff, or other Senior Officials. The Executive Agent who should be the tip of the canine spear, remains under the abject belief that they are only responsible for producing and procuring dogs. We know how that will end … failure to meet the demand for MWD capability, again.

I promised that kid…and God knows over the years, I've forgotten his name…but the bottom line was that he had a problem that leaders needed to correct. If it affected this young man and his dog, it would affect every working dog team that deployed. I thanked him and promised that I'd get it fixed. I was not going to let this kid down.

Chapter 7 – Canine Force Flow

Armed with this information, I did a deep dive into the transportation flow of MWDs, and here's what resulted. Initially, Dover AFB was assigned as the cross-channel port of embarkation (POE) for MWDs. Joint Service MWDs located in CONUS would coordinate home station travel to Dover. However, the handler faced tremendous challenges just to get from their home station to Dover.

The MWD Team was a Transient Orphan

There was no in-transit support. No one had anticipated the influx of so many teams. I thought of the young handler I'd met in Iraq and felt his pain. The problem got worse when the Air Force informed TRANSCOM that Dover was going to have major renovations to the runways. This would curtail Dover as a cross-channel transit site. A new Air Force base would have to be designated to move our MWD teams' cross channel to Kuwait. That would mean routing changes and amendments to deployment orders.

Our first stop was our guardian angel, Master Sergeant (MSG) Heather Dempsey, CENTCOM J4 Deployment Movement Planner. We presented the problem to Heather and in the snap of your fingers, she had a course of action and was already working the phones to make it happen. We knew that the change in MWD transit point had to be an East Coast AFB, but which one? There were several that had the capability for cross-channel missions. Within a few days, with MSG Dempsey leading the way, she miraculously "smoozed" general officer approvals, and McGuire AFB in New Jersey became the new MWD movement site. Through her efforts, the force flow cycle and deployment orders for all MWDs were modified to have dog teams' transit through McGuire AFB. Simultaneously, my job was to establish a CENTCOM reception-liaison team to be our CENTCOM MWD command and control.

That liaison solution came in the person of Senior Master Sergeant (SMSGT) Michael Coons. Mike was cut from the same mold as MSG Dempsey. Dedicated, mission first people always! As the senior enlisted leader for the Joint Security Office (JSO), he managed the manning document. He knew how to work the wickets to bring on additional personnel. For the most part, once you are in the position that Mike was, you've already demonstrated your competency…and ability to "smooze." Mike had the respect of seniors and subordinates.

Mike saw the need to add a liaison position, but that was a challenge. One of the detractors was that additional manning authorizations were normally requested for augmentees to serve at CENTCOM headquarters at MacDill AFB and forward deployed locations. But Mike, like Heather, were doers. He "maneuvered the system" and found a way to make this happen. And just like Heather, with a few phone calls…and concurrence from our division chief, in a few days, Mike got authorization for the assignment of a CENTCOM MWD liaison NCO at McGuire. The dark days of our MWD teams being out on a limb, lost in the transportation void were finally seeing a fix in this first step by creating the "McGuire Hub."

Enter MSGT Reggie Smith!

Reggie is built like a bull and solid as a rock. He's stern looking, but as gentle as a lamb. I don't think that…regardless of the situation or how difficult…I ever heard him raise his voice or get angry. He always had a smile on his face. He always started his conversation off calmly with "Sir" and then delivered his ever so persuasive opinion, and once I saw it Reggie's way, we'd get on with the mission.

Reggie was about to enter the lion's den. As the "sole" CENTCOM representative, he would be looked on as CENTCOM. We had lots to do before we got Reggie settled in at McGuire. But armed with Reggie's impervious, tough, and unwavering demeanor, we mapped out the mission and set in place a course of action to support these joint service MWD teams. Reggie authored the CENTCOM McGuire Hub SOP, which was included in MWD mobilization orders. It directed the following:

    1. Ninety days prior to arriving at the McGuire Hub, MWD handlers will receive reporting instructions.

    2. Upon reporting to the McGuire Hub, you will be considered deployed to CENTCOM AOR. You will follow all deployment guidance (in particular, General Order 1B (no drinking)).

    3. The Veterinarian at McGuire/Dix will service your animals. The McGuire Vet will issue the necessary ten-day travel health certificates.

    4. The Air Force kennels at McGuire will provide training facilities: the obedience course and explosive/narcotics training aids. Handlers are expected to continue odor recognition, bite, and obedience training during temporary assignment at the hub.

    5. The Hub will coordinate a car service for transportation from Philadelphia International Airport to McGuire AFB.

    6. The Hub NCO will coordinate onward travel with the Air Force cross-channel flight team.

    7. MWD teams will be under operational control of the CENCOM Liaison NCO (LNO).

    8. CENTCOM LNO serves as the conduit to ensure that all administrative actions for the MWD teams were accomplished before departure.

    9. The LNO will maintain accountability of all MWD teams via the force tracking numbers.

    10. MWD Hub NCO will notify gaining units of the tail numbers/FTN's of departing MWD teams and scheduled arrival in theater.

With formal approvals from TRANSCOM and Air Mobility Command provided by MSG Dempsey, Reggie and I headed off to McGuire to jump start the Hub. While enthusiastic, we knew one thing. Reggie wasn't coming back to CENTCOM. He'd be on his own to do the "meets and greets," with all the Air Force elements that would support the mission, i.e., Mission Support Group, Aerial Port, Security Forces, and Kennels. The success of the hub rested on his shoulders. As we boarded the plane for Philadelphia, we were on the precipice of a new dawn. I knew at the time that Reggie would hit the ground running as we were already tracking teams that had arrived in New Jersey staying in local area hotels, attempting to get cross-channel transportation.

MSGT Smith was in overdrive from the time we exited the plane. As we moved through Philadelphia airport, he was engaging, getting information, and coordinating with TSA. As we exited the airport, he was inquiring and confirming car service to streamline the process of getting handlers to McGuire. I knew that these MWD teams were going to be well taken care of by one of the most dedicated NCOs I've worked with. The next step was to get to McGuire, roll up our sleeves and get to work coordinating with our Air Mobility Command partners who would be the "cogs" of the Hub.

When we departed Tampa, we "assumed" ... and everyone knows how that adage goes when you assume…you make an ass of you and me. We believed approval notifications had been made by TRANSCOM (the senior headquarters directing global movements) to McGuire. As protocol dictates our first stop was to report to the Mission Support Group Commander, (Base Commander). This is a standard military protocol to conduct introductions but more importantly to establish command relationships. Doing this at the onset avoids any roadblocks for operations to go smoothly. But what always seemed to be lurking in the wings was the unpredictability of Murphy's Law. Murphy was nipping at our heels.

When Reggie and I arrived at McGuire, we followed protocols and reported to the office of the Mission Group Commander, a Colonel, who would be our host. Assuming that this would just be a formality. Unfortunately, when we went into his office, with the intention of thanking him for his support, the reception we received was not what we expected.

With good reason, the Mission Support Group (MSG) Commander was miffed at our presence. He was even more perturbed when we informed him that we were reporting to establish the CENTCOM MWD Liaison operations effective immediately. That led to some tense moments as he had no clue we were coming and was offended that we were on "His Installation" without authorization. We squirmed in our seats, but we maintained our professionalism and decorum. There was a glitch in communications that needed to be resolved as soon as possible (ASAP).

Reggie and I were asked to wait outside. We were apprehensive that this would not turn out well and we would end up having to engage our Stars. Fortunately, this was just a miscommunication, and after a few phone calls to Air Mobility Command, it was fixed. Our "Host," who was now gracious and supportive, promised all necessary support to get us up and running. First nightmare avoided. Relieved and grateful for the renewed momentum of what we had come there to do, we forged ahead, starting the process of knocking on doors and announcing our presence…then asking for favors.

We were not the kings of this castle. We were guests. We followed the established installation rules, procedures, and protocols. It would be the Air Force with their transportation experience and expertise that would be directing this production. We were under the Operational Control of the Air Force Mission Support Group, as a tenant unit. We relied on our AF partners. The AF Security Forces partners who granted our dogs permission to train in their kennels. The Aerial Port Squadron would be the conduit for configuring and transporting all the canine equipment. Air Force installation transportation provided ground vehicles to get our teams to the flight line, veterinarian, or exchange. But most importantly, the Aerial Terminal schedulers who were overwhelmingly kind, worked their scheduling magic and always found a seat for our MWD teams. McGuire and the entire Air Force team were angels on our shoulders.

The saving grace was Reggie. This is where NCOs shine…and with Reggie's bubbling personality, he went right to work. Reggie could charm the eye teeth from a snake. He gained trust and support from the Air Force team. As an Air Force Master Sergeant, he wore the Air Force uniform, understood Air Force culture, and knew the language. We had no command authority at McGuire. All we could do was ask. But in the true spirit of our military, every unit stepped up to the plate. The Air Force adopted us and became a partner in the success of the CENTCOM MWD mission.

I don't think the Air Force really knew the dramatic impact that they had on keeping our troops safe. This "Hub" was truly life and death. We eliminated transportation delays and missing arrival dates by 99.9 percent (mechanical difficulties were not factored as a point of failure). Active duty, Reservist, government civilian, and contractor—everyone rallied around the leash! It was a labor of love to "Take Care of the MWDs." The McGuire Hub changed the playing field. This operation tracked, provided necessary healthcare, and streamlined the canine deployment process. McGuire never received the credit that they deserved. CENTCOM was finally controlling our canine resources. "In Canis Confidorus" or loosely translated "In Dogs We Trust." We were starting to fulfill the promises I made to that young handler back in Iraq.

Operationally, we were Bare Ass Naked! We had nothing. No computers, no office space, no office supplies, no official phones, no government cell phone, no vehicles, no clue yet how many teams were in transit to McGuire. We needed to get set up immediately, and that meant that very day.

First stop, the 87th Security Forces Squadron, the base "Cop Squadron." They had MWDs in their squadron and we hoped they would have pity…and take us under their wing. We dropped in unexpectedly. And to coin a phrase, we probably "blindsided" them. But that wasn't a hindrance. Once they knew what the mission was, they rolled out the red carpet…and like the rest of the McGuire team, they gave us a resounding, "How can we help?"

I was not astonished. I was in pure amazement how the 87th, without hesitation, was willing to give us the shirt off their backs to help. The 87th Chief Master Sergeant, who in the Air Force is a God, started the ball rolling. Within a matter of a few hours, they converted their conference room into our temporary office space. They set up a computer, and government phone. We were in business!

Reflections

As fate had it, the 87th also shared their building with their Air Force reserve counterpart, the 514th AF Security Forces Squadron, or NYPD South. This was an emotional moment, as the 514th had special meaning for me. The 514th Security Forces Squadron was part of the 514th Air Mobility Wing, U.S. Air Force Reserve. This units' previous designation was the 514th Troop Carrier Wing, the unit my father served in from the end of WWII to 1961 at Mitchel Field AFB in Garden City, NY, and then until his retirement in 1968 at McGuire. My dad loved the 514th.

Growing up in Levittown, with the entire community of WWII veterans, the military ethic was instilled in me at a very young age. Respect, commitment, dedication, and selfless service were all part of the daily routine. In the old days of TV, my dad would wake up before 6 am to exercise and I sometimes joined him. I was his little "Battle Buddy." In those days (late 50s early 60s), TV went off the air at midnight and returned at 6 am the following day. When the TV was going off the air, and again when it was coming back on, every channel (all five of them) filled the screen with a flowing American Flag and the Star-Spangled Banner played. I learned to respect and honor our nation's flag. To this day, when I hear our National Anthem, I come to attention, salute, and render honors.

There must have been at least five neighbors who were also in the 514th. It was part of the fabric of our community. I was so proud to see my dad in his uniform with all his ribbons. In the early to mid-60s, the Reserves did their mission. It was not uncommon to see dads bringing their kids to weekend drill. I was always excited when my dad brought me to Mitchel Field as his "Aide." I credit this upbringing and learning to love the tenets of military service as the catalyst for me to pursue a military career. My Dad retired in 1968 as a Senior Master Sergeant, after twenty-eight years of service in WWII, Korea, and was on the pendulum of being called up for the Viet Nam conflict. The reunion with the 514th brought back many fond memories. I'd completed the military circle of life.

Not surprisingly, McGuire is approximately 60 miles south of New York City. As a reserve "Cop" Squadron, it was not surprising that most of the unit were civilian police officers. The NYPD had been hit hard on 9/11. Many of New York City's finest became or were members of the reserves and National Guard. NYC has always been supportive of their reservists, giving their civil servants thirty days annual military leave. The proximity to the city and opportunity to earn another retirement prior to 9/11 was hard to pass up. But now it meant an opportunity to fight those who had attacked NYC.

The 514th AFSF Squadron projected a NYC culture and atmosphere. I felt as if I were back on the streets of Brooklyn. We quickly bonded and the 514th, just as the NYPD protects and serves NYC, looked out for their newly adopted "Rookies." Senior Master Sergeant Ralph Tomeo, just like on the job, was the one who got things done. He was the "mother hen," the one who took care of his "Officers," and he made sure we wanted for nothing. If there was a problem, Ralph would fix it. Ralph went on to become a Chief Master Sergeant. Years later, we ran into each other at the NYPD Counter-Terrorism Bureau. Still serving. But of all the priorities tasked to Ralph, the most important was making sure that during my visits, there were bagels and Bialy's…with a smear…upon my arrival. NYC heaven!

Within five days, we were up and running. Reggie implemented the HUB SOP and had already coordinated three teams for onward movement. Within forty-eight hours, the world knew that Reggie was the CENTCOM Liaison…and the phone didn't stop. Reggie, with our spectacular Air Force partners, processed MWD teams as a one man show. Reggie was on top of getting MWD equipment (500 lbs of support equipment) palletized with the Aerial Port team. He got dogs to the vets for cursory physicals and health certificates. Yes, there were several dogs that were medically unfit for deployment, which meant that we coordinated a replacement team. Identifying dogs that were medically non-deployable at the Hub, allowed Reggie to reach back so we could coordinate a replacement team with minimal delay, rather than detect the medical condition after the dog arrived in theater and was unable to perform its mission.

Reggie also eliminated kids having to pay out of pocket expenses to stay at off base hotels. Shortly after his arrival, Reggie coordinated teams, handler and dog to be billeted in base lodging. Reggie became cook and bottle washer. He validated deployment orders, resolved finance problems, then personally brought and picked up MWD teams at the air terminal. For those teams going forward, he'd get them checked in and once confirmed, he sent the MWD Team manifest and tail numbers to Kuwait. On the reverse side, for dog teams returning from overseas through McGuire, Reggie was the travel agent getting them booked for flights back to their home stations.

For almost nine months, Reggie accomplished this mission non-stop. He was superman, no other explanation. He was able to leap tall buildings in a single bound and was faster than a speeding bullet. But you can't be on call 24/7 without a break. Reggie didn't have a set shift. Teams arrived and departed at all hours of the day and night…and he was there for every movement. To be honest, I don't think I saw him in uniform during those first nine months. He was always in civvies, just coming off or getting ready to work another mission. Looking haggard, but never cross or cranky. He always had that huge welcoming smile, but even he knew he was tired and needed some relief.

Budgeting is always an issue. At the time that the McGuire Hub was initiated, funds to pay for Reggie's off base long-term lodging and car was only approved because of the criticality of the mission. Fortunately, we turned to Senior Coons again, who reached into his bag of tricks and attained a temporary manning position. The position would have to be attained from a local asset, as per diem would not be authorized. What we needed was an administrative NCO, not a dog person. Reggie "smoozed" with Ralph and we were able to attain a super young airman. Just what the doctor ordered. They worked hand in glove and thank goodness they did because the popularity of the Hub was growing.

By word of mouth, the Hub became the transit point for all dogs transiting overseas. We had therapy dogs, three letter government agency dogs, and contract working dogs (another story), just show up. The Hub team never turned anyone away and worked with them to either get them on a "grey Tail" (Air Force Aircraft) or the Air Terminal to make other travel arrangements. Of the dogs for the dogs.

McGuire Hub Epitaph

By the time I returned to Tampa, the hub was up and running. Thanks to the dedication of the entire Air Force team at McGuire, MSG Dempsey, and Senior Coons, the "lost dog syndrome" was eliminated and we attained MWD transportation transparency. We maintained accountability and never again lost a dog in transit. We reduced the delays of transiting MWD teams into the AOR from seventy-eight percent to zero percent. At the height of the War, we moved between fifty and seventy-five dogs to and from the theater every month. The hub was a critical part of the MWD program, and its creation saved lives.

MSGT Smith led the Hub for almost eighteen months before his orders were up. Reggie led the way. We passed the mission on to the 87th Security Forces Squadron who easily stepped in and continued to maintain the forward momentum. It was the personification of "unity of effort." Reggie came off orders and returned to Tampa. However, his story doesn't end there.

With the multiple duties I assumed as the Chief of the Law Enforcement Branch, Mike Coons, our manpower guru attained a contract deputy position…and we hired Reggie. We continued to work together, not only managing the canine program, but also the Customs and Border Protection Inspection program, until I returned to retired status in October 2012. We're still friends today. My brother in arms and brother for life! I am forever grateful and indebted to him…not only for his unbelievable feat of establishing the McGuire Hub…but for his friendship, and always having my "six."

Chapter 8 – MWD Hub McGuire AFB – Coming Home

The McGuire Hub was personal for me, with my Dad having served in the 514th. But that story requires some explanation. My Dad and I were very close until my mom passed away in 1963. When he remarried, our relationship fell apart. We were never close again. In June 1972, I left the house vowing never to return. The good memories of the days of the 514th were all but erased.

However, fate plays tricks on us all. Somehow, while I was in Iraq in 2005, my dad's neighbor in Homassasa, Florida, reached out to me. Before I deployed, I had sent my father a letter, letting him know I was deploying. I must have put my email address on the letter, fate, because there was no way he would have been able to find me.

I was shocked. Apparently, my stepmother had passed several years before and my father—now in his early nineties—was living alone. Worse, he was infirmed and on the verge of losing his house.

This must have been in August 2005. I'd been asked to remain on active duty, with an assignment at U.S. Central Command, Tampa Fl. The die was cast. The next several years would be contentious. There was no love, no identification of this man as my father. But there was that military connection. My Dad was a WWII and Korean War veteran…and we always take care of our troops. I had to do a mental assessment to take 'father' out of the equation and look at him as a troop who needed care. It was the only way I could deal with my anger and emotions. It had been thirty-three years since we had last seen each other.

When my wife and I reported for duty at CENTCOM, we made the arrangements…and then made the trek for the reunion in Homosassa. There was a tired old soldier, bedridden, desperate, needing help. My anger turned into a desire to help him…and I did. From 2005 to 2009, we made sure that he got the best care through the Veterans Administration, eventually putting him in a VA foster care program. The foster program is an assisted living arrangement, where veterans live in a caretaker's home. He wanted for nothing. Medical appointments, meals, entertainment were all provided. What a godsend.

The love of his life was his WWII B-26 Bomber, a Martin Marauder, whose nose art was "My Gal". He wished more than anything to be able to see a B-26. There were several aviation museums in Florida, and I reached out to CENTCOM Public Affairs and enlisted their help. To our amazement, they found a B-26 exhibit at Fantasy of Flight Aviation Museum in Polk County. Fantasy of Flight was thrilled to be able to give my dad a special VIP tour. When we got there and approached the B-26 his eyes lit up as he saw the true love of his life. He'd been a crew chief and almost immediately, he started spouting off the mechanical settings of the Rolls Royce 2400 engine. I was proud to have been able to take him there. At that moment, I realized that we were closely bonded…as soldiers. I was glad that I'd been called to serve in the last years of his life, for being the catalyst for him to reflect on his military service and see his beloved B-26 once more.

As the years passed, his health declined, but he was happy. On July 30, 2009, he passed quietly. His wish was to be cremated and buried with his wife, in their family plot on Long Island. He was a veteran, a true American hero. I love him for his dedicated service and the example that he set that put me on my own path. I wanted to give him the honors he deserved but had no clue how to pull an Air Force Honor detail together.

Reggie, unbeknownst to me, set that all in motion. The 514th provided a funeral detail for his internment. I took on the responsibility of the stressful task of the process for opening the grave for the internment at St. Charles Cemetery in Farmingdale NY. I had to provide copies of death certificates and the day of his cremation in Florida selected an appropriate urn where Sergeant Vargus' remains would be laid to rest. I selected a porcelain one, plain, but adorned with what we referred to as the Air Force "Wing Nut." The current air force symbol adopted from the Army Air Corp of the 1930–40s.

We went up to New York in early August. I was in awe of how many people came to pay their respects. When we were escorted to the grave site, I was speechless. The 514th Honor Guard…flag team, firing squad, and bugler…stood crisply at attention in their Air Force ceremonial uniforms. I had to stop and catch my breath. Reggie had worked this miracle. I was astonished, but so very proud that my father's final honors could be rendered by his beloved 514th.

The funeral director placed the urn on a ceremonial platform. The guest lined up in front of the urn in a loose formation. When it was my turn to present his eulogy, I hesitated. Should I focus on what he wasn't or what he was? I chose to honor a fellow service member. He served twenty-eight years, in war and peace. He dedicated himself to the nation, and the Air Force. That was his legacy. As I concluded my eulogy, I did an about face, called everyone to attention and slowly raised my hand in his final salute.

The members of the flag detail marched deliberately to their positions. On that day, they were the sentinels in this garden of stone. Simultaneously, the firing squad marched into position, preparing to render the 21-gun salute. The firing squad detail NCOIC's voice crackled, "Detail ready…aim…fire." Three times the volleys rang out. When he called the team to present arms, the flag detail crisply snapped the flag and started the process of meticulously folding it, thirteen folds, with six white stars displayed, while TAPS sounded.

I fought to keep my emotions in check. I wanted this final honor for my father. Once folded, the flag team NCOIC maneuvered the flag so that the butt (the long end) would be presented to me. But I wasn't accepting the flag. I asked that the flag be presented to my cousins Steve and Barbara. They were part of "Uncle Art's" life. He had been a part of their family. Their Dad, my Uncle Steve, was also a WWII veteran. I knew they would accept and honor this flag. I was not a son…and he was not a father. I felt it appropriate to present the flag to them. I had completed my mission.

This was a special honor for me as well. The honor guard from McGuire had traveled three hours in traffic, to honor a fellow Airman. They came not only to honor my father, but because an NCO (Reggie) was taking care of his Colonel. This detail from the 514th will always hold a place of high esteem in my heart. They gave my father, Senior Master Sergeant Arthur A. Vargus, a fitting soldier send off to his final duty station.

Chapter 9 – MWD Hub Kuwait

From time to time, a miracle arises, and you never look a gift horse in the mouth. I'd heard the old adage, "Does lightning actually strike the same place twice?" Well, for the canine program, it was about to strike a second time. McGuire's founding father was MSGT Reggie Smith. The founding father of the Kuwait Hub was a young Air Force Master Sergeant, Brian Umbaugh. He was the "god father" of what would become the CENTCOM MWD Hub at Al Salem Air Base in Kuwait. His efforts, like Reggie's, are lost to the canine ages. Only those who lived those dark days still treasure the legacy of Brian, Reggie and their predecessors. MWD heroes, doing their job, the MWD "transit chain," totally run by NCOs.

We knew that when MWDs arrived at Kuwait International Airport, it was assumed they were part of the unit on the passenger manifest. Troop handlers knew that when units arrived, their direction was to get them processed and moved to one of three acclimation training camps. Everyone. At that point, the disastrous misdirection of MWD teams started. MWD teams are certified and pre-trained. Acclimation training is not required! They needed to be on a direct track to their supporting units, but their transit and logistics flow was disrupted. No temporary kennels, no veterinary medical care. Lost in the sauce, at acclimation camps, some up to forty-five days without resupply.

I needed to look at how to take disaster and mirror the McGuire Hub. I had the opportunity to travel to Kuwait, where I spent time with the ARCENT Provost Marshal and forward deployed MWD Program Manager, SFC Dennis Ford. Several dog teams had just been bounced back from one of the acclimation camps, and Dennis was working the wickets to get them manifested to their units in Iraq. Their stories were all the same. They'd been left relatively on their own to fend for themselves and their dogs. It was depressing. I again thought of the young handler I had met in Iraq and the promise I made. When Army dog teams were rounded up they were sent to the Army transit facility at Camp Doha. This was a tent city tasked with moving thousands of Army personnel.

The handlers waiting to go forward told me that they were being helped by the kennels at the adjacent Ali As Salem Air Base. To my amazement, the Air Force had a system in place that ensured accountability for Air Force MWD teams on Air Expeditionary Force (AEF) rotations. They were tracked, accounted for, provided veterinary care, provided temporary lodging, one stop shopping. This was an Air Force only hub. They had their shit together!

The major Air Force Security Forces Command in Iraq was the 732nd AFSF Group. They had oversight security at all Air Bases in Iraq. Ali As Salem was their transit hub. The 732nd assigned a Liaison NCO there to manage Air Force personnel going and returning from Iraq, including MWD rotations. It was a well-oiled machine.

Al Salem Air Base Security Forces also had assigned MWDs and an operational kennel. Supporting the base kennels was an Army Veterinary Team (Veterinarian-Vet Tech).

The concept of "One team, one fight" for our four-legged heroes was starting to come into focus. What we needed was to consolidate Joint services MWD movements and make this the second leg of their journey. McGuire – Kuwait – Iraq/Afghanistan. We needed a CENTCOM MWD Hub. How could we cut and paste this efficient process.

MSGT Umbaugh…like all the NCOs I had the honor to serve with…was another methodical, focused, motivated young hard charger. Everyone was younger than me, so they were all "young hard chargers." One evening, I stopped by the 732nd Liaison Trailer, introduced myself, and found out that

Brian was not only the 732nd Liaison NCO, but an AFSF MWD Handler. We immediately hit it off. He understood the problem and we sat up most of the night brainstorming. Canine leaders are all the same, we seem to be cut from the same canine mold.

As we talked, he had the exact same concept…to centrally process all incoming dogs. Take the 732nd Liaison template and create a replica for the U.S. CENTCOM Joint MWD Transit Hub. I spent the next two days with MSGT Umbaugh learning the process. He showed me how they were notified by the Movement controllers at Kuwait International Airport when a 732nd airman/dog team arrived. He explained how troops were transported to Al Salem AB, signed into the unit, and then processed for transit forward. The dog teams saw the Vets for a cursory medical. Everything was done under Brian's watchful eye. Once flight arrangements were coordinated, he escorted the "Defenders" to the air head for their flights to BIAP (Baghdad International Airport – HQ of the 732nd).

The 732nd had this system down pat. However, one of the immediate concerns was how to establish a notification system. How would the Kuwait International Airport reception team notify the "hub" of arriving Joint Service Teams? MSGT Umbaugh and I traveled to KIA to see what needed to be done to establish that function. To this day, the simplicity of why Officers should stay out of NCO business never fails me.

The Air Force Aerial Port was the unit that received personnel and equipment arriving in Kuwait. It was the first stop. From there, the practice was, that all Army personnel moved to Camp Doha. The Aerial Port would contact the 732nd Liaison with notification of arrivals and Brian would coordinate transportation for the incoming airman to Al Salem. MSGT Umbaugh knew all the key players, introducing me to the Chief Master Sergeant in charge. Within minutes, they solved the problem. The solution? Brian scribbled his name and contact number on a piece of paper and the chief put it on the notification board. The Aerial Port personnel were to call the 732nd Liaison when Joint Service dogs came off the plane. From that very second, the CENTCOM Hub was born. It didn't take long for this "new" process to be tested.

Proof of principal was tested that very evening. MSGT Umbaugh received a call from the Aerial Port identifying an inbound cross channel flight with an Army MWD team. By then, I was MSGT Umbaugh's Tonto and where he went, I followed.

Around one a.m., we drove the 732nd van down to the Aerial Port. The aircraft had just landed and there were several Army troop handlers waiting to separate the Army personnel and get them on buses to Camp Doha, Buehring, Virginia or Camp Arifjahn. I informed the Army reception teams, that from now on, MWD teams would be picked up by the AF 732nd Liaison team and brought directly to Al Salem Air Base. That didn't go over well, but after a phone call, it was understood that the Air Force would process *all* MWD teams.

I wish I would have been able to capture the look on the MWD Handler's face, a young looking SSG and his "shep" as he came down the aircraft stairs. Waiting for him was an Air Force MSGT and Army Lieutenant Colonel. When we announced his name and said we were here to pick him up, he stopped dead in his tracks. I think he believed he was being arrested!

As we grabbed all his equipment from the pallet and boarded the van, he asked why we were there to pick him up. I explained that this was a test run for what would probably become the new centrally managed MWD transit Hub. He was ecstatic and started to lament about his previous deployment. Just like the young MWD handler I'd met in Iraq, he told me of the nightmare trying to obtain transportation from his unit in CONUS, being pushed from hither to yon for days before being able to "hitch" a ride to Iraq. His testimony solidified that establishing this MWD transit Hub in Kuwait was mission essential.

Once we got back to Salem AB, MSGT Umbaugh assigned him to a tent, provided him an information sheet on the DFAC (Dining facility), the location of the Veterinarian, admin instructions for reporting to the 732nd LNO so he could be processed for onward movement. Brian was amazing, he single-handedly reversed a process that was impacting canine operations. He was taking care of MWD teams.

That evening this ad hoc proof of concept set the stage for success. The next phase was to get stakeholders buy in to officially sanction a joint services hub. We worked on the details of how to staff and find resources necessary to get it up and running. At that point we were again, as we were at McGuire, just about bare ass naked!

The following morning, I made the rounds, speaking with the AFSF Squadron Commander, the Air Base Commander, and the Air Force Central Command (AFCENT) A3, who was the senior Air Force Headquarters for air operations in the Middle East. For the Hub to succeed we needed "buy in" from all stakeholders.

But this took time. This was the beginning of Brian's individual superhuman efforts to assume accountability for every single MWD team that would transit to and from Kuwait. For the interim, HE was the HUB. He and Reggie became very close. But here's where teamwork and caring for troops shines through. Without the stupidity of bureaucratic games, here is where troops pull together to take care of their military brothers and sisters.

Brian aimed to be the pinnacle, trying to work 24/7 to stay on top of the K9 force flow. He was a superman, but even superman needs to sleep. However, once the word spread that he needed someone to pick up incoming MWD teams, airmen volunteered left and right. Every organization on the base joined the effort! These unsung heroes did more than their share…at all hours of the day and night…airman volunteered to make the two-hour trip down and back to pick up joint service dog teams. This was their mission too. Hats off to the men and women of the Air Force. Their selfless service helped save lives.

Let me take a quick step back to summarize. I hope that you're getting the picture. The wonderful books that have been written about our hero dogs tell the "good news" stories of the unbelievable missions our canines have performed in incredibly dangerous situations. Did any of you even fathom the pieces of the puzzle that had to be put together to support our MWDs before, during and post deployment? These HUBS were born from an idea; that idea had to be documented, coordinated, resourced, and executed. I hope that this illustrates just how much time and effort was required to move our dogs. You should be asking yourself…what the "f"? I say that because as I've pointed out, the process is already out of sight/out of mind…and we are going to repeat this debacle come the next major war. Guaranteed.

And please, my intent is not to detract from the influence and support of the Officer Corps at all levels of command. My counterparts and superiors at CENTCOM, ARCENT, AFCENT, our services, all supported the MWD program. But the worker bees, the kids who moved the teams, the leaders who brought the Hubs to life, were NCOs, and the credit for this magnificent initiative belongs to them. The accolades belong to them, too. I just knew how to put the pieces of the puzzle together.

Putting the Pieces of the Puzzle Together

For six months, Brian and his volunteers continued this ad hoc effort. During that time, we laid out the concept of operation, engaged AFCENT (Air Force Central), who owned all AF assets in CENTCOM. Who would own the Hub? How many people would be required to staff a 24/7 operation? Who would provide the vehicles and temporary quarters? Would we need to construct a facility? Or would we need to double up with one of the AF units? All valid questions. The Hub was unique, as the Operational Order directing its establishment was going to be a CENTCOM Operations Order! CENTCOM would be the owner operator, manned jointly by Air Force and Army personnel.

As we started the process, the first challenge, no new manning positions. Manning would have to come out of hide from existing units of the Army at Camp Arifjahn and Air Force personnel from the Security Forces Squadron at Al Salem AB. Then Murphy jumped in for fun. Our initial plan was to build a Hub, but the Kuwaiti government instituted a policy of no new construction, on military bases. There was some push back and apprehension, but we eventually worked through each of the organizational concerns, putting the final touches on the U.S. Central Command Operations Order. With the staffing process complete, the chain of command established, our Senior Commanders from AFCENT and ARCENT in concurrence, and the OPORD approved by the CENTCOM J3, the MWD Hub at Al Salem AB was scheduled to launch in April 2008.

The Operational Order directed all MWDs to transit through McGuire AFB with onward movement to the Al Salem HUB. The Kuwait Hub would be jointly staffed (two Army, two AF), with the Air Force assuming the leadership responsibility (The Air Force would manifest teams on aircraft moving to Iraq and Afghanistan). The Army troops would be fully integrated. Not only would they share troop handler responsibilities, driving, admin, but they would also have to learn the Air Force airframe scheduling system. The Hub was going to be truly "Purple" (Joint)

The Kuwait Hub would operate 24/7. Like their short stay at the McGuire Hub, MWD teams would be under the command of the Hub Air Force NCO in charge. As the summer of 2008 approached, we were rotating approximately 100 dog teams in and out of theater a month. The OPTEMPO was at fever pitch.

MSGT Umbaugh was the founding father of the Kuwait Hub. His mantra being, "failure is not an option." Brian did the near impossible, maintaining continuity for many thankful MWD teams. He rotated back to the states just prior to the Hub kickoff. He will always have the admiration of the MWD teams that he ensured were not lost in the sauce. And Brian, you will always have my admiration for vaulting this hurdle. I'm proud to have worked with you as my mentor, and proud to call you a friend.

I flew to Kuwait to collaborate with the team to verify our resources and facilities. However, Murphy showed up first. As the team of stakeholders came together on a miserably hot day in Kuwait, we were confident that the I'd dotted the I's and crossed the T's. But the launching of the Hub was still in limbo the restrictions on new construction had not yet been resolved. We were in jeopardy of having to curtail the mission.

Red Horse – The Air Force Santa Clause

As is always the case, leave it up to the NCOs to find the solution, or "manipulate the system." They'll find a solution to get the mission done. Sometimes twisting and interpreting regulations askew from what a layperson might consider the letter of the law, but they will get the job done. I'd never heard of "Red Horse," a specialized Air Force Engineer Deployable Heavy Operational Repair Squadron. They really were miracle workers and the saviors of the Hub!

While we sat around the table trying to figure out how to fix the facilities problem, enter the Red Horse Chief Master Sergeant and several of his team. We rehashed the issue and the perplexity of the situation and presented our dilemma to the Chief. Without hesitation he momentarily conferred with his team and then informed us that this would not be a problem. Apparently, he "knew" where there were several trailers that were not being used, and he asked if they would fit the requirement. He already knew that under the restrictions imposed by the Kuwaiti's, these trailers could not have a permanent foundation. Not a problem.

The Chief was no nonsense and down to earth. Tell him what you want done and get the hell out of his way! He set the tone with the stakeholders, asking who had the design, how many kennel runs were needed, electric, plumbing/running water, sanitation run off, prep-area, and where we wanted them. The selected Hub NCOIC, another stellar AFSF NCO, presented his rough schematic. That being confirmed, the Red Horse Chief then asked if all the Officers could clear the room so he could get down to work. The "officers," given our marching orders, tucked our tails between our legs and dutifully exited, leaving the mission to the experts.

Thus, began what would go down in the annals of canine history as the miracle at Ali As Salem. It was an unbelievable feat, and I'm glad the officers who had nothing to offer and would have been a thorn in the side of the Chief stayed far away, allowing the Chief and his team to create this symphony of building perfection. The choreography of moving parts was a sight to behold. My guess was…that with all that had to be put into place, from scratch, moving and setting the trailers on the earthen hard stand, installing the internal utilities, all in 130-degree heat… would take at least a week to complete.

Red Horse proved me totally wrong. As I went about my business and the day came to an end, I went to my quarters and settled in for the night. What I saw in the morning blew my mind and forever established my indebtedness and total awe to the Red Horse Miracle workers.

In the morning after breakfast, I wandered over to the Hub Admin trailer (previously the 732^{d} LNO). I couldn't believe what I saw. The Hub was complete! As I walked toward what looked like brand-new double-wide trailers, I kept a safe distance, not wanting to be told to get the hell away from the Red Horse airman! The Chief was nowhere to be seen.

Red Horse had outdone themselves. They had not only set the trailers in place, but they also built six kennel runs out of chain link fencing in each trailer. When I flipped on the lights, they worked. Each trailer had a functional air conditioning unit, a prep-area with running water, and a spicket hookup to hose down the kennel runs. It was an dream come true. How could all of this have been accomplished in just over twenty-four hours? Simple, it was Red Horse and that's what they did best, "the impossible." When you think of the contributions of our military service members, there is something unique and humble about these amazing men and women. When you express your thanks for going above and beyond, when you compliment them for being stellar performers, the response you get is universal, "I was just doing my job." The Red Horse response was the same, they were just doing their job.

But what they did, in just over twenty-four hours, was allow the smooth transition of MWDs to our forward deployed units. They ensured MWD command and control. Their resilience effectively eliminated "lost dog syndrome." We now had theater MWD lines of communication. From McGuire to Kuwait, to the gaining units in Iraq, Afghanistan, and our other locations. One hundred percent accountability. Just doing their job!

We met our mark. From that great day in April 2008, through the reduction of major operations in theater, the Hubs were the canine lifeline. We can only hope and pray that the efforts to create this movement template will become part of canine archives, readily available for our next major conflict.

Chapter 10 – Contract Working Dogs

After we instituted the CENCTOM Admin Instructions and reporting procedures, we still had one category of canine that we needed to get a handle on...Contract Working Dogs. We were unaware of how many contract working dogs were supporting multiple-US agencies and U.S. contractors throughout Iraq. We had no clue if they could detect explosives or if they were certified or how they were being maintained or their health status. There was a disconnect even though...under the CENTCOM instructions...we had command and control for ALL dogs, period.

There were multiple issues. First, Theater Joint contract command approved contracts. They were a team of military contract specialists, trained solely to review contract statements of work. Ensuring financial continuity as stewards of government funds and ensuring the legality of the contract prior to award. What could go wrong with a canine contract that lacked canine subject matter oversight? Everything! There was no interface with the Theater or CENTCOM MWD Program managers! Contracts were being awarded, period. Working with military canine experts was crucial to ensure the fidelity of performance, certification, and veterinary health (work cycles, environmental issues, vaccinations). Without this input, Contract Command approved canine contracts in a bubble, allowing contractors to operate with total anonymity.

In 2009, the Provost Marshal in Multinational Command – Iraq requested our Program Managers and theater Kennel masters perform a deep dive to get a handle on CWDs. There had been too many reports of dogs...working in his backyard...who were unable to detect explosives. And this situation was becoming common knowledge. Some twenty canine contractors from South Africa, the UK, and Ukraine were operating under Joint Contract Command approved contracts.

Re-Setting the Chess Board

CENTCOM coordinated a "hit" team to go to Iraq and evaluate every CWD team. We mustered a seasoned team of Army Kennel Masters and Program Managers from across the Army and sent them to Iraq for forty-five days. They had one mission: validate the detection deliverables in every contract. Previously, without the CENTCOM MWD policy, contractors refused military program managers admittance to their facilities. Denying access to the military to perform canine due diligence. They were now on notice. Meet compliance standards or be terminated. A lot was at stake for these contractors. Unchecked, these companies received millions of dollars to provide canine force protection. But the big question was, were their dogs trained to detect explosives? Or were they bringing in untrained dogs to look nice at the gates, unable to detect anything and still reaping millions in US dollars?

Fortunately, the Sheriff was back in town...and for some of our less than stellar canine contractors, what we believed was happening was...appallingly...taking place:

- Handlers had no clue how to read the behavior of their dogs or how to introduce dogs to a search pattern for explosives.
- Dogs missed explosive training aids.
- Dogs did not have the proper health certificates (vaccinations) and were working with restrictive medical conditions.
- Teams were working well outside of feasible work schedules.
- Work schedules were being dummied.
- Team training was not being conducted.

- Contractors did not have standardized explosive/narcotics training aids.
- Dog certifications had been expired for months.

We terminated contracts, established corrective action plans and performed revalidations. We also proactively worked with their teams under the watchful eye of our MWD program managers. Don't get me wrong, this was not a blood bath. We needed the contract working dogs. They were essential. What we did was restore contract integrity. We turned the page with inclusion of the MWD subject matter experts. We modified the CENTCOM policy, the modification directed all contracts be vetted by a canine program manager and veterinary corps officer.

We needed to get contracting command to buy in. Reversing a slippery slope, changing paradigms and lines of communication. We became conduits for contracting officers overwhelmed with an avalanche of canine contracts. Working 24/7, these contract officers were attempting to perform insurmountable tasks. Literally, in Iraq and Afghanistan, there were hundreds of contracts to provide services and rebuild national infrastructure. Our MWD teams and Veterinarians evoked a canine contract standard template. Under our contract oversight, the MWD/Vet team wrote the statement of work. Every contract was validated by an MWD Subject Matter expert and Veterinary Officer.

We ensured that contracts included MWD program managers open access through announced and unannounced visits. Not only to perform validations and health and welfare assessments, but to work with their dog teams to improve proficiency. A win win. We took back control, and at any point, if a contract dog failed detection validation or a dog's health was found to have been neglected, we had the oversight to correct the issue. It was another burden thrust on our already stretched program managers and kennel masters. Yet in our business, there is no room for error. The difference of a healthy, properly-trained dog capable of detecting explosives —or not—was life or death. Where we ran afoul in this new oversight role, is that our MWD professionals had to quickly become familiar with Contract management. Adding that to their already bursting responsibilities was the assumption of being appointed CORs (Contract Officers Representatives). The job of MWD handlers — trainers, Kennel Masters, and Program Managers was to provide trained certified MWDs to support the military. Our MWD troops were neither prepared nor trained in contract management. It was not a task presented in the MWD Kennel Masters Course or the military education system.

As a COR you are charged by the government to be a steward of government funds. You are responsible for verifying services performed. You become legally accountable to approve monthly payments. Being a COR was another job not known to the public, another mission that our great MWD leaders executed… adapt and overcome. Unfortunately, there was a glitch in the COR training. MWD leaders designated to assume COR responsibilities could not take the COR course prior to deployment. The Joint Contracting Command would not issue appointment orders to incoming Kennel Masters and Program Managers until they arrived in theater. The period of appointment was for the duration of their "boots on ground." But as always, our NCOs took on the responsibility and executed the mission admirably.

This was one of the lessons learned that has since been implemented into the Army MWD leader education program. But not realized at any of the COCOMs. Right now, there is a huge gap in our war planning variables. First MWDs are not identified in any of the COCOM's war plans. Second, contract support is not identified for any capability. Early on contractors became a vital element of the warfighting team. We framed the best course of utilization; MWD's support maneuver units outside the wire; contract dogs provide force protection at static forward operating bases. The joint team of MWD/CWD will be an enduring requirement, a vital part of the theater canine footprint.

Our MWD and Veterinarians work miracles. I was fortunate at the COCOM as a Lieutenant Colonel, I had the rank to influence and direct the program, through the living policy and my ability to interface with senior leaders. The EA identifies contract working dogs (CWD) in their MWD policy guidance. But the EA has no concept of contract dog operations, contract management, training and oversight necessary to execute a CWD program. How can they advise senior leaders and the SECDEF.

Chapter 11 – Cadaver Dogs

But there was a contract working dog that the military was unable to provide oversight— cadaver dogs. Cadaver dogs would play a significant role in the positive identification of deceased High Value Targets (HVTs). Tactics, Techniques and Procedures using drone strikes was one method to take out HVTs. How does an infantry squad or Special Forces team positively identify human remains that can be analyzed in a forensic lab? You only got one shot at this. Minus a positive DNA test, the next best capability for positive identification was a cadaver dog.

Cadaver dogs can work through rubble and detect bodies under water. The military has no cadaver dogs or the technical name "Human Remains Dogs (HRDs)." That capability exists solely in the private sector. Therefore, a contract requirement had to be executed. Two issues prevailed in the execution of a Cadaver Dog Contract. One, the military had no cadaver certification authority and no expertise in validating a cadaver dog's statement of work. How would the military oversee any contract? We would be at the behest of the contractor and accept the cadaver dog national civilian standards, normally endorsed by FEMA.

Second, training—what would they train on? Where would they get their "human remains" training aids? How would the training aid be accounted for? What is the work cycle of a human remains dog? Who would provide medical support—private or military veterinarians? Who would serve as the Contract Officers representative (COR)? The eleven-page Statement of Work titled, "Tracker/Cadaver Dog Support for Personnel Recovery," did not address any of the above concerns.

The only guidance was captured in the General Section, paragraph 1.2 that stated, "the contractor must possess thorough knowledge and understanding of the Army Missions in Iraq and Afghanistan and have extensive experience in the recovery of human remains under exigent circumstances, inclement weather, and in difficult/varying terrain." The remainder of the contract presented requirements for Handler qualifications, scheduled and unscheduled absences, performance of service during crisis declared by National Command Authority, identification cards, Clothing and Equipment—and a conglomeration of administrative requirements. Standard language in all contracts.

We were entering a canine "grey area" but needed the capability. Contract command approved a contract for, "Two human remains dogs stationed in Iraq at the Kennels at Camp Victory for administration only." The annual cost was $1.5M. The contract was a firm fixed-price meaning—that regardless if the cadaver dogs were used or not—the annual payment to the contractor, AK9, was set in stone. One of the concerns we had throughout the war was that in accordance with DoD guidance, working dog contracts needed to be vetted by the Air Force as the Executive Agent. But the EA was far removed from any engagements with the CENTCOM MWD Program. CENTCOM filled the EA gap.

Unfortunately, training aids were abundantly available. Kennel master—HRD teams were able to obtain them from the Medical Treatment Facility (MTF). HRDs trained on discarded bandages primarily, but trainers sometimes were able to introduce their dogs to real human cadaver parts. The HRD teams trained and trained and trained. Their proficiency was unquestioned. While the regulation prohibits intermingling contract dogs with military dogs, we sort of looked the other way and provided space for the cadaver dogs in the military kennel. We learned about this special capability…and it helped us learn how they were trained and assess the long-range requirements for possible utilization in the MWD inventory.

However, by the completion of the first performance period (1 year), the cadaver dogs remained virtually unused. I believe they confirmed two High Value Targets (HVTs). A cost-benefit analysis demonstrated no actual value for spending $1.5M. But as a special commodity, Commanders wanted the dogs on standby.

The contract became contentious when U.S. Forces Afghanistan (USFOR-A) requested a cadaver dog. The statement of work read that the contract supported Iraq and Afghanistan. However, the cadaver dogs' statement of work verified that the dogs would be stationed in Iraq. To move AK9 cadaver dogs from Iraq to Afghanistan required a contract modification. Contracting Command determined this was outside of the Statement of Work and that a financial modification was required. This change resulted in an invoice to move the dog from Iraq to Afghanistan of approximately $500K per animal. Fortunately, this never happened because this mission was curtailed. Yet we learned another lesson on the importance of verifying every facet of a contract.

Within the next year, it was determined that the contract was no longer required. Yet AK9 proved their weight in gold – an honest contractor that placed the mission first, above their profit margin, never looking to bleed the government. True partners and committed to accomplishing the mission. We'll see them again.

The AFGHAN Contract

After the Presidential mandated withdrawal from Iraq in December 2011, the priority of effort returned to Afghanistan. The Taliban was attempting their own "surge" to regain control of the country. CENTCOM did a mission analysis—and as requested by the USFOR-A Commander, CENTCOM re-missioned units from Iraq that still had up to six months remaining on their deployment.

As part of that dynamic mission analysis, the U.S. Forces-Afghanistan Provost Marshal and MWD Program Manager conducted a theater needs assessment. Based on the tactical situation, and the increased casualties from IEDs, it was apparent that Military Working Dogs were spread too thin. MWDs were overtasked, supporting troops outside the wire patrols, and performing force protection at Forward Operating Bases (FOBS).

The USFOR-A Provost Marshal and MWD Program manager were strategically brilliant! Switch out all Military Working Dogs from the static mission of guarding fixed installations and replace them with contract working dogs. The concept would release MWDs to concentrate on their primary mission outside the wire finding IEDs, keeping our troops safe.

Using contract working dogs to support troops on foreign soil was prohibited. From a legal standpoint, there was no Status of Forces Agreement (SOFA) between the Islamic Republic of Afghanistan and NATO allies. SOFA is an agreement that grants authority to perform military operations on their sovereign soil. Basically, an invitation. Legally contractors were not military personnel. Contractors performing outside the protected confines of a Forward Operating Base were considered "mercenaries." If a contractor committed a violation of Afghan law outside the wire, they could be arrested and tried under Afghan Law. Following those legal protocols contract working dogs were exempt from prosecution while working on a coalition operating base. But the undertaking was substantial. The contract would require 300 dogs.

This course of action was a huge step outside the traditional MWD box and a supply and demand issue. The reality, future enduring operations will need to identify contract working dogs as part of the canine deployable inventory. The cost…let's just say LOTTO, it was in the millions.

However, cost wasn't the issue. It was developing the contract statement of work, providing performance and medical requirements, and developing regional lines of communication. Due to the numbers of dogs Army vets were unsure if they could support veterinary care, and the costs for procedures and medical prophylaxis was expected to be exorbitant. The contract called for veterinary care to be provided by the contractor. Private veterinarians were not too keen to go to Afghanistan, the closest they would come was Dubai. Immediate emergency care would still have to be provided by Army veterinarians. To alleviate the costs on Army veterinarians, they developed a reimbursement schedule. The contractor would be required monthly to write a check to reimburse the US government for medical treatment. Another lesson learned.

The size and management of the contract was outside the capability of the Joint Services Contract Command. Afghanistan. And so, the Army Contract Command in Flint Michigan, 12,000 miles from Afghanistan, was selected to be the Contract Officer (KO).

USFOR-A wrote a solid performance work statement. Working closely with them we concurred with the template for excellence in contract working dog certification and utilization. Army Regulation 190-12 sets forth standards for certification and utilization for Army MWDs. These standards were direct and stringent. CENTCOM adopted AR 190-12 as the standard performance template for all CENTCOM canine contracts. Army veterinarians' regulations for the health and welfare of MWD's would also serve as the medical template for CWD care and maintenance.

This contract reinforced that the MWD chain of command would forevermore require deploying canine managers to be well versed in contract management. As Afghanistan was its own theater of operation, the contract was called the Theater Wide Internal Security Services (TWISS). The numbers of contract dogs at some bases would be twice the numbers of military working dogs. Our kennel masters would be engaged more with contract validations, training and reviewing monthly vouchers than with MWD's. The biggest challenge was that regulations forbade CWDs being co-mingled with MWDs. Yes, we overlook the regulation with Cadaver dogs in Iraq, but that was only two dogs. Not the case with TWISS. How and who would provide separate contract facilities and how would that cost be figured in the contract?

TWISS also changed how we managed contract dog pre-deployment preparation. In 2009 when we sent a "hit" team to Iraq, we were playing catch up. This contract was not going to be the CWD debacle in Iraq. From the date of the contract award, CENTCOM would direct pre-deployment screening, processing, and certification. HME imprinting would be conducted in the U.S. where practicable. CENTCOM would have oversight of the hiring process. We performed quality control, verifying a slice of personnel records. Army veterinarians reviewed every CWD veterinary record. This change in contract working dog pre-deployment preparation was almost identical to the MWD pre-deployment process. A new chapter in contract oversight. Contractors arrived in the theater fully vetted, ready to go to work. Quality control was a force multiplier.

Implementing these changes did not come without a bill payer. It required manpower, primarily MWD certification authorities. While CENTCOM didn't own any troops, the services (Army, Navy, Marine Corps and Air Force) were part of our task organization. Unfortunately, we had no tasking authority. That hurdle had to be overcome. That meant relying on relationships and yes…favors. I was blessed and extremely fortunate that that problem was easily overcome as my requests for support were never turned down. Thankfully the generosity of our service Program Managers sending certification authorities to certify the contract dogs at Hill Country Kennels in San Antonio. Every "volunteer" certification authority paid for their travel expenses. The result of this MWD team effort, a consistent certification rate of 98.5 percent.

This effort was not without the collaboration and cooperation of the contractor, American K9. They were in business to turn a profit, but their entire leadership team were mission focused. The mission was just as important as their profit margin. TWISS was a success because there was unity of effort between customer and contractor.

The team of CENTCOM, USFOR-A, Service Certification Authorities, Contracting Command and American K9 worked hand in glove to achieve this tremendous effort. The true proponent for evaluating working dog contracts was the responsibility of the Executive Agent. But during the war no contracts ever came across the EA desk. Is the EA executing their responsibility? What pro-active steps are they taking within the Joint Service MWD Committee to prioritize or develop a course of instruction for the services on contract operations. Did the EA brief or will they be expected to brief the SECDEF or Joint Chiefs of Staff on contract working dogs? Will the EA assume strategic responsibility for DoD policy for contract working dog employment and utilization for domestic and overseas contingency operations?

What is the current policy for contract working dogs? Has each combatant command performed a theater wide assessment of their full spectrum canine needs? The reality, doubtful to not at all. You'd probably have a better chance of pulling a rabbit from a hat. Canine lessons learned have passed into the ages.

Training Coalition Partners - Travesty of Errors

The main mission of the war once coalition forces removed the Taliban and Saddam Hussein from power was the reconstruction and rebuilding effort. We flooded the country with training teams to rebuild their security forces, reconstruct their infrastructure, and rebuild their national sovereignty, or to coin the phrase "With Liberty and Justice for All". How would that be achieved. How would we infuse unity after centuries of cultural power struggles between the Shia and Sunni's.

Under the Multinational Security Transition Command - Iraq or MNSTIC-I (Min sticky) Police Academies, and Police Training Teams (PTT's) were stood up and deployed throughout the country to train the Iraqi police. What about dogs? What about working to ramp up canine detection capability. That task was passed to the Army Military Police MWD community to introduce basic canine training.

Iraqi canine training was initially established at the Baghdad Police Academy. MP dog handlers would work with the Iraqi's and their dogs, working on basic obedience, and detection. Along with our veterinarians, grooming standards and sanitation and care of the dogs. We never knew where the dogs came from, and many dogs were unsuitable for training.

But the attitude and culture of our Muslim brethren was that dogs were dirty animals and unclean. Initially this created multiple concerns, as our goal was to ensure that these dogs were healthy and serviceable to perform a detection mission. Having MP's provide foundational training with this cultural negativity towards dogs proved to be one of the two major concerns. Eventually leading to prohibiting MP's from supporting Host Nation Canine training. The Iraqi's had no handler selection process. Second, their failure to adhere to the instructions of the MP MWD handlers, and third they had no respect for the dogs. In short order, many of the dogs were abused, several found dead in their kennels.

This was also detrimental to U.S. troops. We were taking MWD handlers out of the fight. Pulling them from supporting units on combat patrols, leaving a gap in MWD operational protection.

The final straw, the Iraqi government held the MP Training Teams accountable for negligence in the death of their dogs. As such the Iraqi government was demanding reparations of $10K per dog. Unbelievable, but of course to maintain good relationships with our Iraqi partners we ended up paying the reparations. We also immediately revised the CENTCOM MWD Policy prohibiting MPs from training Iraqi dog teams. We would serve in an "Advise and Assist" role. If the Iraqi's wanted trained dogs, they would have to source a private contractor.

That's what they did. Nationwide K9 in 2008 negotiated a contract with the Iraqi Minister of the Interior for $1,530,317 to train master trainers and seventy-five dogs. Whether the contract was fulfilled and met the performance work statement was strictly between the Iraqi government and the contractor. CENTCOM was officially finished with host nation training liability.

Hopefully this lesson learned will be a cut and paste for future contingency operations, keeping our MWD assets where they belong… in the fight.

Chapter 12 – Force Multipliers – Traditional to Special

By the beginning of 2005, the prosecution of the war in Afghanistan was shifting from a conventional to Special Forces "heavy" mission. Rotational Special Forces (SOF) groups were tasked with small unit covert operations along with our CIA brethren. The focus that SOF would be successful at identifying Taliban and Al Qaeda leaders and dismantling their inner circle. Their mission was to gather intel and prosecute High Value Targets (HVTs). Special Forces teams…Operational Detachment Alpha (ODA), 12-man self-sufficient units…were the foundation of the special operations missions in Afghanistan. Stealthily deployed by Army Special Aviation Detachment Aircraft, they were extremely mobile, maintaining the element of surprise. Our ODAs…just as our conventional forces…were all reliant on one critical force multiplier. Dogs!

SF units employed Multi-Purpose Canines or MPCs. These dogs were specially trained to operate off leash out to one hundred yards. They were not trained by the Department of Defense MWD School, but by private contractors under special operations funding and SOF statement of work. MPCs were specially trained for bite work, tracking, and off leash explosive detection. However, as SOF's Operational Tempo increased dramatically in 2006, the demand for MPCs quickly overwhelmed the contractor's ability to produce a surge of MPC's. And the mission wasn't going to slow down anytime soon. So how would SOF adapt and overcome to get sufficient four-legged force multipliers to drive this SECDEF priority mission?

Robbing from Peter to pay SOF

The Israeli's and British adopted the concept of off-leash dogs. This was the birth of the Specialized Search Dogs (SSDs). These dogs were nonaggressive, single purpose explosive detection canines trained to search at distances of up to seventy-five meters.

SSDs provided an additional buffer (extended distance) to protect our troops from an IED blast. SSDs expanded the search areas utilizing the dogs mobility allowing larger areas to be cleared more quickly than a methodical search with a dog on a six to thirty-foot leash. The SSD detection mission: find different types of ammunition, explosives, IEDs, weapons caches, and the dreaded home-made explosives, whether they be freshly buried or hidden in a five-hundred-pound plastic barrel.

The Army also saw the significance of adapting the SSD. The Deputy Secretary of Defense established the "Integrated Process Team (IPT) for Defeating Improvised Explosive Devices (IED)" in an Action Memorandum dated 17 July 2004. In his responding memorandum issued two months later, U.S. Army MG Fred Robinson, Chairman of the IPT, approved the Army to purchase of thirty-nine Specialized Search Dogs (SSD) to deploy to the USCENTCOM Area of Responsibility (AOR).

Both the Army and Marines desperately needed this capability to support combat operations against our Al Qaeda, Taliban, and Mehdi Army adversaries. The capability, while sanctioned in Israel and the UK, would have to be approved as a program of record by the Air Force as the Executive Agent. Under pressure to project this new platform, the Air Force approved the capability. This new program of record would be established at Lackland AFB under the auspices of the 341st Training Squadron. The drawback—as the program had not been adopted by all the services—SSD instructors would have to be provided by the Marine Corps and the Army. They adjusted their annual canine procurement numbers, including the additional Marine and Army SSDs. But this did have an impact on Army and Marine MWD readiness. Pulling instructor manpower from the Army and the Marine Corps broke qualified teams. The argument was for having experienced civilian instructors vs. military manpower.

The SSDs would be directly employed by the Marine Expeditionary Forces or MEFs. The Army split their SSD requirements between Military Police and Army Engineers. For good reason, the Army MP SSDs supported dismounted patrols (foot) executing searches in buildings, vehicles, open areas and roadways. Army Engineer SSDs would be solely utilized to clear major routes (roadways) to keep the lines of communication open. Keeping major thoroughfares clear supporting vehicle convoys moving troops and supplies free of explosive threats. The ninety-three-day course commenced at Lackland and the first SSDs were certified and deployed by early 2006.

The Winds of War

At that time, the theater was populated by Marine Expeditionary Forces deployed to Anbar province and seven Army Brigade Combat Teams. But there was a major utilization shortfall. The Marines SSDs were embedded, meaning they were built into the unit as part of their "force package". The Marines built force packaged units that included their mission enablers: aviation, logistics, engineer, dogs and any other special capability. They were self-sufficient. The Army, to this day, uses the augmentee concept for dogs. Army dogs, to this day, are requested and deployed independently. They are not built into a force package. One of the positive aspects of a force package was that force multipliers could not be touched. In force management language they were "fenced". As Army MWDs were not built into a force package unit, they were fair game to be re-missioned, which is exactly what happened.

As the Army and Marines were graduating their first SSDs, the winds of war shifted. The National Command Authority direction to Special Operations Command (SOCOM) was to defeat the Taliban in Afghanistan, once and for all. Whatever type of enablers were required from conventional units, priority of effort would be to direct or redirect as was the case, to Special Operations command. One of these identified assets was "Off leash Dogs."

Sometime early in March 2006, just as I had been handed the torch as the Chief of Force Management, I received a phone call from the Military Police Force Management Officer at the Joint Forces Command, informing me that effective immediately I was to remission all the army MP SSDs to Special Operations units in Afghanistan. That this order came from the top and there was no negotiation. Wow! I had no clue how to proceed. I was still getting to know all the stakeholders and where they fit in this big puzzle.

The paper drill was no problem. We just amended their deployments to the Special Operations Groups, switching their force tracking numbers (FTNs). That didn't affect their mandated latest arrival dates. That was the easy part. There were serious issues that this re-assignment created. One, when MWD teams arrived in theater (in this case Afghanistan), conventional dogs received a thirty-day acclimation period before they were released to their supporting maneuver units. They adapted to the environment, running through detections scenario-validations introduced to theater specific explosives.

That didn't happen in the SOF world. MP SSD teams were NOT provided a thirty-day acclimation. Upon arrival, our teams would in-process, receive their theater specific equipment, get a mission briefing and deploy the following day. These SSD Teams were not Special Forces trained. They were conventional Military Police MWD teams. They were not special forces qualified and the physical fitness standards for dog handlers were established by each service policy. SOF personnel's physical and endurance requirements were those of an Olympian. The demands of their mission mandated a better level of physical fitness. SOF expected a SOF capable trained team. In the beginning, this created a problem with unfair expectations and an immediate mistrust of the ability of our MP teams to support the demanding missions of SOF.

Then there was the gender issue. SOF at the time was restricted to male soldiers. There are no gender restrictions for MWD handlers. Male or female, regardless of gender, there is no distinction of a certified team. They go forth and execute the detection missions to protect troops, military equipment, and dignitary security. There has never been a restriction and to be honest every one of our MWDs handlers are champions in my book. Yet this was a huge problem, in some cases, female handlers were either prohibited from going on mission, or requested to be transferred to a conventional unit. Jesus the issues of physical limitations and field sanitary conditions were embedded in SOF mind set. But from my fox hole, economy of effort and availability of MWD's was the driving force. We couldn't afford to be in the business of disputing MWD handlers' gender.

We were robbing Peter to pay Paul as it was trying to meet the SOF MWD gap. Each of these young men and women were proficient and efficient. Fortunately, the need for dog teams overrode the desire to retain an all-male ODA. Once our SOF were exposed to the professionalism and resilience of our female handlers the walls of mistrust came tumbling down. We made sure that we included in the notes section of MWD teams assigned to SOF that MWD handlers would not be gender restrictive. We sent the best of the best. Period.

However, the biggest shortfall was pre-deployment training. SOF was spot on with their mistrust of the capabilities of our MWD teams, especially when we ran out of Army SSDs and had to change the platform to "an in lieu of" requirement. Robbing Army, Navy, and Air Force Patrol Explosive Detection Dogs (not off leash) to fill the void. Eventually we exhausted the SSDs and had to change the requirement cross leveling Army, Navy and Air Force non off leash teams to support SOF. We had no choice. The cupboard was bare.

Every MWD team was deployed based on their service rotation policy. The Army and Navy MWD teams for a year, Air Force teams initially six months then eventually modified to a year. Special Forces Groups were six months, but their deployment cycle commenced when they started pre-deployment training at home station. Then their drawdown upon return to home station approximately thirty days. All of this amounted to approximately nine months in total. Our conventional MWD teams' deployment clock started when they arrived in the theater. There was no MWD transition. Conventional MWD teams came in cold. There was no continuity with the ODA.

What should have been the standard deployment cycle would have been service conventional teams begin their deployment with the ODA during their stateside pre-deployment ramp up. A full SOF rotation. Learn their mission, integrate with the teams, and determine if the team has the METL to perform a SOF rotation. See if the MWD teams were able to meet the SOF physical and operational demands. If a change was needed, we would have sufficient time to change a team, rather than identify an issue with a team once they reported in Afghanistan. Under the SOF mission, our MWD handlers were not just handlers. They were fully integrated into the ODA. They adapted by meeting both very demanding physical and mental challenging missions.

How do we address this in the future? Simple. Get MP teams "ODA qualified" and have that SF MWD Qualification annotated on their record. Establish a cadre of conventional teams from all the services. Once qualified, promote training integration. This would be a step forward for both the SOF and conventional communities to be able to provide identified MWD force enablers for future needs. Asymmetric warfare is a critical task not taught to MWD teams, yet we know that asymmetric, rather than traditional missions will face service traditional dog teams. The seven "Ps" apply: Previous Prior Planning Prevents "PISS POOR PERFORMANCE." The question is…will the SOF MWD lesson learned become a change in doctrine or remain a skeptical concept and a bridge too far?

Resolution and Challenges

The outcome of our joint conventional/SOF missions revealed some very interesting facts. The breakout of dogs was a 50/50 split of Special Forces MPCs and Army, Navy and AF dog teams. MPCs and conventional dogs complimented each other. The kennel master and trainer were NOT Special Forces troops, they were Army MPs. In the end, the SOF MWD mission was an operational success due to the principle of unity of effort. We found that the SOF mission formula will be an enduring requirement. We provided a needed capability. We also broke down a few barriers along the way, improving interoperability to support SOF in future contingency and "Nation Building" missions. It was the capability not the gender of the handler or service uniform.

MWDs will be utilized with priority based on need and a formula of a basis of allocation to support conventional forces, and a slice to be determined by our SOF organizations. Those enablers will come from the joint services. We also learned that we need to look at increasing wartime MWD populations as a planning factor. Not wait until the first shot is fired! And there will be additional requirements. Will we be prepared? Sadly, I think not. We can only hope that this lesson learned will not be forgotten and as quoted in a recently published Naval War College Thesis on Military Working Dogs (Babbitt Naval War College Thesis, 2013) the truth probably lies in her brief commentary:

"Today's placement of the DoD MWD program within the Office of the Undersecretary of Defense for Intelligence (OUSD(I)) construct removes it from the eyes of the operators, effectively burying it in a community that cannot and does not most effectively articulate its operational protection value to Combatant Commanders."

The Army MP School, the proponents for Military Police doctrine foresaw the future. Through a joint manpower study, with the Office of Provost Marshal General and Army Special Operations Command, an Army MWD Advisor was added to the Army Special Operations Command. This bold and positive step allows the MPs to be better integrated…assisting in identifying MWD force enablers, and the tactics, techniques and procedures conventional teams will need to train to be fully interoperable with our SOF brothers in arms.

Chapter 13 – Tactical Explosive and Improvised Explosive Detection Dogs

By 2010, casualties were mounting in Afghanistan due to improvised explosive devices. These peroxide-based explosives were causing more causalities than enemy indirect fire. The services were stretched to the limit supporting both conventional and SOF forces. Field Commanders were calling for the one proven capability that they knew could mitigate this threat, the only force multipliers that could protect their troops...MWDs.

In the summer of 2010, the Commander of Combined Task Force Strike (2ᵈ Brigade/101ˢᵗ Airborne Division) submitted an Operational Needs Statement (ONS) for Improvised Explosive Detection Dogs. Positing that from 2008 to 2010 there had been an eighty-three percent increase in the use of IEDs resulting in thirty-seven percent increase in casualties. The request supported a recommendation to train non-traditional Improvised Explosive Detection Dog teams, using infantry soldiers pulled from the ranks. A modified MWD course. Providing partially trained teams, but something had to be done.

The ONS submitted through the Commander, U.S. Forces Afghanistan to the Chief of Staff of the Army (G3/5/7) requested 121 additional non-traditional dog teams. Consequentially, the Marine Corps had also identified IEDs as the number one threat, and the need to provide more explosive dogs in the fight. The Marines were developing a non-traditional Improvised Explosive Detection Dog program (IEDD). It was a temporary program; a stop gap. Standing up this non-traditional IEDD program was not meant to replace a trained/certified Military Police MWD team.

The Marine Corps program was quickly approved and funded. They established a contract with K2 Solutions in Pinehurst, NC. Like the Army and Marine Specialized Search Dogs, the Marine program would provide "off leash" single-purpose explosive detection dogs. The Marine Program was well established and functional. Selected from infantry units and based on their adaptability and desire to work with a dog, Marine handlers were well versed in infantry tactics...and they understood how canine teams would be integrated and employed as part of the mission. IEDD teams were embedded with deploying Marine units. Once IEDD teams completed training at K2, they returned to their units and attended pre-deployment training to integrate into the infantry, maintaining cohesion and continuity. They deployed as part of the team, executed their mission...and returned together as a unified team.

Based on the request from the 101st Airborne Division, the Army staff reviewed and approved the request. However, the Army established their own separate program, the Tactical Explosive Detector Dog (TEDD.) Regardless that their TEDD program mirrored the Marine program, they believed that TEDD training would provide additional detection skills to best support Army units. The Army deemed it necessary to establish a separate contract for TEDD. Contract oversight was performed by the Office of Provost Marshal General. My job at the time as the CENTCOM MWD Program Manager was to track the dogs, include them in our monthly theater statistics, and track their performance.

In the summer of 2010, I conducted a CENTCOM MWD conference at our headquarters in Tampa. The services, Army veterinarians, and theater MWD program managers attended. This was one of several that we conducted during my tenure. The goal was to provide a theater canine mission overview. And, as a priority topic of discussion, have the Marines and Army provide briefs on their non-traditional IEDD and TEDD programs.

We provided the most recent monthly canine statistics, and our projections for MWD strategic deployments for the next two rotations. CENTCOM continued reliance on the AF and Navy to support conventional maneuver units and Special Forces. We were stretched thin. We were eager to hear the Army and Marine plans for executing their non-traditional MWD programs. From the briefings, we realized that the Marine Corps and Army programs mirrored each other. Both were temporary, training infantrymen on basic detection. Their employment—once trained—was to deploy with their unit for a single rotation and provide counter-IED support. Upon completion of their tours, their dogs would be returned to the contractor…and these temporary handlers went back to their infantry duties.

What I gleaned from their briefings was that the Marine Corps program was organized, effective, and efficient. They were already providing infantry units with additional canine force multipliers. From CENTCOM perspective…as these two programs were parallel in nature…we believed that this it would be in the best interest of continuity for the Army to piggyback on the existing Marine IEDD program. Consolidate effort, and finances.

However, the Army stated that their program would provide additional skills, enhancing their team detection capabilities. Their TEDD program would be executed through an Army contract. The contractor would procure two hundred dogs, mostly Labs, and train them at their facility. The plan was to provide each deploying Army Brigade Combat Team with twenty embedded dog teams. The Army Director of Operations published an Execution Order directing the Army to stand up the TEDD program. The Office of Provost Marshal General worked diligently to frame the statement of work and at the beginning of 2011, the TEDD training contract was approved as a sole sourced contract to Von Liche Kennels in Indiana. What was unique is that neither the contract nor funding was approved through the Army Contract Command. However, there were other avenues through "special" funding lanes that I'm sure were used as the catalyst to provide the startup appropriations.

The Marine selection process was formidable. Liking dogs and having a pet is nothing compared to training a dog for war. The handler must accept the daily responsibility for the health, welfare, and maintenance of an animal. The selection criteria worked well…and yes, having a basic knowledge of a dog was but one of the selection criteria. Being physically fit and recommended by their commander as a motivated hard charger with no disciplinary actions were also criteria. Corporals (E4s) were the desired handlers. The test of the program facing K2 was matching handlers with dogs.

The Army process differed dramatically from the Marines. There were basically no criteria. The Army selection process at the beginning was controlled chaos. Commanders selected anyone who stepped up to the plate and volunteered. The Army had clerks, cooks, truck drivers, and yes soldiers who were "voluntold."

TEDD training was not coordinated. The selected TEDD handlers were sent to Von Liche during the same time their BCT were conducting pre-deployment training at the National Training Center and Joint Readiness Training Center. Pre-deployment training overlapped, and TEDD teams were still in training and being certified up to two months after their BCT's deployed to theater.

This led to the second problem, lack of leadership. When the twenty troops were selected to deploy to Von Liche for training, there was no Brigade Leadership. These "kids" were on their own. For the most part, they were well-disciplined and attempted to learn their jobs. However, for many the "call of the wild" was too much and they either quit or failed the course. During the TEDD program, the typical attrition rate from the basic course was between fifty to seventy percent. What that meant was that between thirty to fifty percent (seven to ten) handlers who volunteered completed the course and made it through "basic handlers' training." Then they went to Yuma Proving Grounds for certification. Of those numbers, anywhere from twenty percent to forty percent failed certification. Of the original twenty handlers approximately five to seven made it through certification. And then there were further eliminations when TEDD teams were unable to complete theater acclimation successfully. Yet, senior leaders refused to see the forest through the trees, supporting a program that was no more than a government waste of funds.

The third problem of oversight that arose during the period of the contract. The Office of Provost Marshal General, acting as the Contract Officers representative, never followed contract protocols. VLK had the contract for almost eighteen months—and not once did an Army contract officer's representative conduct a site/COR visit. The contract ran under the radar of Army oversight. It was executed sight unseen. The reality was that the program was broken to begin with and remained broken throughout its existence.

The Army would not accept ownership of the dogs until the team passed certification, so there was an element of checks and balances. The contract to provide twenty trained basically skilled teams never achieved the "Statement of Work." In comparison, the Marine Corps IEDD program, a contract officer representative was permanently assigned to K2. This on-site representative managed and maintained oversight on the training and contract administration. The Army entered blindly into a contract and then placed their entire trust in the integrity and honesty of the contractor. The Marine IEDD program ran effectively with minimal wash-out rates, solely because of solid contract management.

As the program continued, the Army did revise their selection process to closely mirror the Marines, ensuring a skilled hard-charging infantryman. But that did not improve the graduation or certification rates. The Army failed to adjust training to parallel the Brigades' deployment cycle, and TEDDs continued to arrive well after the Brigades were deployed. An estimated seventy percent of government funds for this contract were wasted.

Where Were They Used?

CENTCOM waited anxiously in the hope that these non-traditional teams would help turn the tide and reduce casualties from IEDs. But we also needed to know how these dogs would be tracked, how they would be employed, who was going to be their Canine Chain of Command, how were they going to receive logistical support, and medical care? Would this pose any additional responsibilities for our MWD Kennel Masters? TEDDs and the Marine IEDDs were not tracked like traditional MWD teams. They were embedded. Their unit was supposed to provide the full spectrum of support.

Day of Reckoning

Because Army TEDDs did not deploy with their units, they were sent to a staging area at Bagram Air Base in Afghanistan, where the VLK Field Service Representatives (FSRs) conducted their scenario-based acclimation training. As the CENTCOM MWD Program Manager, I was requested by the U.S. Forces Afghanistan Provost Marshal…to be on hand and observe the first arriving TEDDs. I met and worked closely with the MWD Program Manager for Afghanistan, Navy Senior Chief Petty Officer Ricky Neitzel. He was a methodical canine professional. With more than fifteen years of canine experience in the Navy. Smart and organized, he was unaware of how his life…and those of the Canine Leadership in Afghanistan…was going to be disrupted.

When the first TEDD teams designated to support the 101st Airborne arrived at Bagram, there was no 101st reception team. There was no NCO leadership. The Senior Chief had to quickly pull in MWD Kennel Masters and trainers from around the country to help with the TEDD acclimation process. pulling them off existing missions. These TEDD handlers, many of them on their first deployment, had no clue what was really expected of them. This was a baptism by fire.

The VLK Field Service Representative (FSR) and MWD interface was not good. There were two VLK FSRs to validate more than forty-five TEDD teams. They had no permanent office space and no office supplies. They had one vehicle, a broken-down VW van, to transport the TEDD handlers and their dogs to training. They had no training aids. There was no plan for acclimation. It was all haphazard.

The FSRs were less than stellar performers. It was a mess. Scenario-based training was redundantly slow. Both FSRs were required to run a TEDD handler, one at a time, through training, while the other 44 or so mulled around doing nothing. If this was going to be a deep dive into what was supposed to be a three-year contract, would the TEDD contact be an efficient utilization of the time and effort, spending millions for little or no return?

Murphy was traveling with these teams as once the realization set in that they would be going outside the wire, many of these inexperienced first time deployers lost interest, got frustrated, and just quit.

The 101st assigned a brand-new 2nd LT on his first deployment with zero knowledge of canines as the TEDD Officer in Charge. Each day was a crap shoot. The FSRs had to beg, borrow or steal training aids. The MP kennel masters and trainers that Senior Chief Neitzel consolidated at Bagram, tried to run rough shod where they could, reinforcing basic search techniques and how to read changes in their dogs' behavior. Their feedback was less than promising. What should have been a refresher was not. The KMs and trainers felt like they were teaching basics.

Logistically, TEDDs were supposed to receive support from their Brigade. But the Brigade leadership had no one assigned or any plan to support them. The entire plan went out the window the moment the TEDDs arrived in Afghanistan. TEDDs became totally reliant on the MWD Program Manager and Kennel Masters for all of their needs. The kennel masters shook their heads and took TEDDs under their wing. Not one KM turned their back on any of the TEDDs. "In Canine Confiduris."

One thing was obvious. We couldn't keep our MP leaders and handlers away from their own kennels or their primary missions of providing MWDs to units operating in hostile areas. I envied the dedication of the MP "cadre" that supported this infill. They put in many hours mentoring, but it was obvious that their frustration was reaching a boiling point.

One evening after meeting with the FSR, a group of MPs were talking about the day's events, voicing their frustration. They had had it! Baby-sitting was not part of their mission. Then it came to a head…and I swear if we had an academy award for frustration the academy would go to… a Staff Sergeant then, a Sergeant Major now…went off! "F" this and "F" that, "go F yourself, these f..ing idiots!" "This program is friggin stupid." His mastery of the F word flowed so smoothly. I wish I had made a recording. But he told the story, one that solidified my after-action report to the Commander USCENTCOM. As you can imagine, my report was not complimentary.

TEDD was rushed. It was created with good intent, looking to provide dogs to keep troops safe. Where did the failure lie? Right back to the Air Force, the Executive Agent. None of this would have been necessary if the Air Force had done its job and had a plan to fulfill their DoD charter…to procure and provide dogs for the services. Think about it—the real culprit escaped unscathed.

CENTCOM and JIEDDO also must take some responsibility. Based on the IED threat, funding and expanding the growth of MWD's should have been an integrated priority by JIEDDO. Unfortunately, as I've described the millions provided to JIEDDO focused solely on technology. Dogs were an afterthought, not even considered a counter IED system.

The EA and JIEDDO skirted their responsibility Allowing the services to execute their own independent program split command and control and left a huge gap in the logistical support chain. Non canine senior leaders placed inexperienced "partially trained" troops in life and death situations.

The unfortunate fact…there are still no plans for MWD surge requirements in war plans. The TEDD and IEDD programs thankfully were temporary, three years. Yet the aftermath would haunt the services for years to come. When senior leaders do not trust their subject matter experts, and beat their chests, believing that because of their ranks and positions, they have the tools to make decisions… they become the weak link in the chain. This was no more than an "off the street" endeavor, selecting handlers and giving give them less than one third the training time of a traditional handler. It was unfair to the soldiers, and unfair to the operational units.

I was at odds with the Army. They stood up a program but never went to Afghanistan to perform a program assessment. The Army blamed me personally for the TEDDs that we sent back to the contractor because they failed theater acclimation. But the facts spoke for themselves. Theater program manager reports from Afghanistan, month after month, from the first rotation to the end of the program in December 2013, painted the same picture. No explosive finds, utilization rates were terrible. The overall program find rate for the combined three years of the TEDD contract was approximately .0001 percent. Utilization, that included training time was supposed to be twenty-four hours "weekly. Monthly reports reflected a dismal five to ten hours.

More disheartening commanders boasted of the success of the TEDD program. The truth was ignored. It was the Marine Corps program that demonstrated success. However, the Marines were also realizing that the utilization of their IEDDs was not getting the biggest bang for the buck, and they started to reduce their IEDD footprint in late 2012.

By the beginning of 2012, the Army Office of Provost Marshal General had yet to send an assessment team to Afghanistan. While I had been to the theater three times, the cloud of darkness still hovered over TEDD. I consistently re-enforced my concern and the concerns of our deployed Kennel Master's every time I had the opportunity to brief senior leaders at CENTCOM.

Through persistence, OPMG agreed to send an assessment team. With the Joint Security Office (JSO) as the host, we traveled to Afghanistan in March 2012. CENTCOM sponsored a joint team: Army Staff (OPMG), ATF and DoD Holland Veterinary Hospital. We arrived in Afghanistan and visited kennels at Bagram and Kandahar. We visited with the Brigades, who initially stated that TEDDs provided an added value to their mission. But when confronted with use and numbers of finds, they could not justify their effectiveness. Some Commanders even allowed their TEDDs to become unit mascots or mis-utilized them strictly for base force protection.

Those commanders who did not have confidence in TEDDs reluctantly confessed that if they needed dog teams, they would request MP MWDs rather than use TEDDs. The inquiries were not impressive. However, the Army OPMG team believed that their program was a success. I didn't budge from my position. But as you'll see further in the TEDD saga, this would all come back to bite me in the ass.

Unfortunately, regardless of the glaring facts, Brigade Commanders wanted TEDDs. They belonged to them, and regardless of their performance, regardless of their lack of utilization, the concept of ownership is ninety-nine percent of the law…and, with the Commander being the king of their domain, TEDDs remained.

My time at CENTCOM was ending. By April 2012, the National Defense Strategy began reducing forces in Afghanistan. With the withdrawal of troops from Iraq in 2011, the need for reserve and retiree recall augmentees was ending. As CENTCOM reduced their footprint, reservists, and retiree recalls…many who initially joined the CENTCOM staff in 2004 and became branch and division chiefs…were being released from active duty. The blessing of being allowed to serve my country in a time of war was ending.

Payback

That reality happened in May 2012. My Division Chief informed me that my orders were going to be curtailed and that I would be released from Active Duty and returned to the retired rolls effective 12 October 2020. I was fifty-eight and had already retired from the police department. I aspired to remain in Tampa, hoping for the opportunity to get a job at CENTCOM as a civilian. Unfortunately, that was not in the cards. The military is a unique society. We never really leave. For so many of us, we end our military careers and transition to a similar job in the military as our military skills can often be applied to civilian positions.

I was not any different, applying for civil service jobs concentrating on Florida and the national Capitol region. I submitted numerous applications to multiple federal agencies. I was usually referred to a hiring authority, but not selected. Undeterred, I continued the process, finding a vacancy announcement for a Senior Law Enforcement Management Analyst, that included oversight of the Army MWD Program. The position was in the Office of Provost Marshal General…my nemesis.

I reluctantly applied, knowing that the "institutional knowledge," for as long as I can remember, was the "godfather," Hans. A civilian contractor who had been the Army Canine expert for over ten years. I assumed that even if I applied, Hans would be selected for the position, OPMG taking care of their own, the right thing to do.

I applied for the position and was contacted in mid-October for an interview. I assumed this was a formality and that Hans would get the job. But after two weeks, I was informed that I had been selected for the position. I was shocked. What this meant… that I would inherit the TEDD program, the one I'd argued had no value and should be eliminated. Penance.

I reported for duty in November 2012, and was partnered with a former MWD Program Manager who I'd worked closely with in the past. Jimmy had been the Army Pacific MWD Program Manager. We were "oil and vinegar." Jimmy had been the lead OPMG representative on the visit to Afghanistan and a staunch supporter of TEDD. The monthly utilization reports from Afghanistan were still abhorrent, now reporting zero training hours, zero finds. TEDDs were no longer force multipliers, but force detractors.

Exasperated, I transitioned into the job, and assumed the responsibility of the TEDD Contract Officers representative (COR). The TEDD contract was in the process of being re-bid. Since VLK had executed the contract for the base and first year extension, it was generally assumed that they would be the front runner. To my amazement, they were not. K2 Solutions, the company executing the Marine IEDD contract was selected. This created massive turmoil…and for the first time, I was personally accused of having influenced the selection process because of my open comments and distain for the program while at CENTCOM. Nothing could have been further from the truth. But it didn't end there. After the contract was awarded, Congressional staffers representing VLK's congressional district summoned OPMG to the Hill, to have a "chat." My division chief, Mr. Williams, an articulate, well-versed retired Colonel, and I went to the meeting to clarify concerns. We thought that the meeting was clear and concise. But this was just the beginning of a shit-storm that would last for almost four years.

Government contracts are unique. In this case, the TEDD contract was not directly awarded to K2 solutions. It was awarded to a prime contractor DPMS. K2 was the sub-contractor, performing the statement of work, training the dogs. Contract administration and payments went to DPMS for distribution to the sub. DPMS was the middleman. However, having the prime contractor near the Pentagon allowed for a better interaction…and we established a monthly face-to-face meeting.

The next step was the transition of authority. We were going to do this right. As the Contract Rep, we coordinated a trip to inventory and transfer dogs from VLK to K2. When we arrived at VLK with our K2 transition team, the air was tense. However, to my relief, VLK was professional and supportive. We conducted our serial numbered inventory of each dog, captured all the training records…and I accepted ownership on behalf of the government. Phase One. We still had the contract oversight for all TEDDS currently deployed with at least three BCT rotations in Afghanistan. At the time, we believed that the contract would be extended for a fourth iteration. CENTCOM had not expressed any anticipated drawdown.

For months following transition, utilization rates were non-existent. We asked when the TEDD contract would be curtailed, asked CENTCOM for a transition plan so we could work with contractors to efficiently prepare to draw down the contract and transition the dogs. We continued deploying TEDDs, however, we began developing a transition plan that assumed five courses of action:

1. TEDDs would be assessed and based on medical disposition, they would be reassigned to fill gaps at Army kennels.
2. TEDDs unable to be placed against vacancies in Army Kennels would be offered to the other services.
3. TEDDs unable to be transitioned within the DoD would be offered to Law Enforcement.
4. TEDDs would be offered to former handlers.
5. TEDDs would be offered for adoption to the public.

Crucifixion

On December 13, 2013, in a very short email from the CENTCOM MWD Program Manager, OPMG received notification that the TEDD program was no longer needed effective immediately. We welcomed the news. However, along with the end of CENTCOM's need for TEDDs, a domino effect of actions and decisions, had to be accomplished, with no reaction time. First there were 229 TEDDs in training or deployed. The contract was scheduled to conclude in Feb 2014. The immediate concern, an annual cost saving of $1.3 million. Or should we extend the contract as the Marine Corps did to allow for a period to affect the downsizing and smooth disposition of the dogs.

The decisions by OPMG were cost driven. They decided that the Army would not extend the contract past Feb 14. Through my division chief, we pulled out our draft courses of action. I was directed to implement a strategy to dispose of 229 dogs in approximately six weeks (Feb 14). That was a juggling act! The dogs going through training we could manage, but how do we manage TEDDs whose redeployment dates were not scheduled until August 2014? Technically, they were no longer under contract. How do we affect adoptions for TEDDs no longer owned by the Army?

Jimmy kicked off the transition. He coordinated with the Army Commands MWD Program Managers to go to K2 to evaluate dogs for selection for the Army program. Simultaneously, through OPMG law enforcement contacts with the International Association of Chiefs of Police (IACP), we ramped up the first week in February 2014, to bring in law enforcement agencies and give them the opportunity to adopt the remaining TEDDs for their respective departments.

Enter Mr. Rob Squires. Rob was assigned to Army Forces Command (FORSCOM) as the TEDD program manager. Brigade Combat Teams (BCTs) were FORSCOM units, and Rob assumed the duties of tracking TEDD deployments. I was glad to have him on the team as this was a massive undertaking. Rob was a godsend! Not only did we have approximately 189 dogs to dispose from K2 solutions, but approximately forty TEDDs in Afghanistan.

The first week of February, Rob joined me at K2. Our task, to assist with the processing of TEDDs to law enforcement agencies. Within a week, more than thirteen law enforcement agencies attended the event and selected what I believe were forty-five to forty-seven TEDDs. A good start. We were going through the motions, following the DoD adoption checklist. If a law enforcement agency presented a letter from their department, outlining their need and requesting a military canine, we executed the documentation and the department assumed ownership. The same practice applied not only to the TEDD program, but to all adopted DoD dogs.

Thankfully Rob Squires helped immensely during law enforcement week. He looked at what other courses of action we could take to encourage adoptions to private citizens. He was the catalyst to prepare a rolling accountability spread sheet of the agencies, number of dogs adopted and status of remaining dogs. At the conclusion of that week, we needed to regroup and act on the next phase. We still had close to eighty TEDDs at K2 and forty in Afghanistan. I made a plea to extend the contract for six months to give us some breathing room, but my recommendation was disapproved by OPMG. My marching orders were to meet the deadline, to have all TEDDs disposed by the end of February.

Remember the Executive Agent…the one that failed to provide the surge dog requirements? They are also the final authority for disposition of DoD Dogs. How were they going to support the process? They were going to pass the buck. The EA provided a letter to the Army authorizing them to execute disposition of TEDDs on behalf of the EA. The Army had no Executive Agent assistance, we were on our own. Unfortunately, the short timelines gave us little time to prepare a solid long-range transition.

As word of TEDD adoptions hit social media, former TEDD handlers joined with a social media advocacy group, Justice for TEDDS demanding that former TEDD handlers be given "their dogs." That opened the flood gates of accusations, threats, and misinformation. Of course, if it's on social media it must be true and the public and politicians demanded answers.

There was a fine line between official and unofficial. But the word that was used by the media and elected officials was intentional. Unofficially we looked ahead. And thank goodness for Jimmy's foresight. Jimmy was the TEDD certification authority, and he would go to Yuma Proving Grounds to complete the deployment certifications. During each session, Jimmy advised TEDD handlers if they had an interest in adopting a TEDD when the program ended to provide him their contact information. The list was unofficial. The bottom line…every TEDD handler that went through Jimmy's certification was offered an opportunity "to sign up." The allegations that OPMG took no action was again false information, fitting the political narrative.

Rob worked with Jimmy contacting former TEDD handlers. Almost immediately former TEDD handlers who had not heeded Jimmy's request to provide them their contact information came forward demanding to adopt their dog. Claiming that TEDDs they deployed with for a single nine-month tour were their personal property.

Rob and I were threatened and attacked maliciously on social media. It was a three-year assault by strangers and advocacy groups, led by *Justice for TEDDs* whose sole purpose was to demand the return of TEDDs to their handlers.

Let me reiterate, not *intentional*. OPMG failure was not considering and extending the contract to generate a smooth transition. The Marine Corps extended their contract for a year to repurpose and dispose of their TEDDs. While they had similar accusations made against them, they were minimal. No one from *Justice for TEDDs* wanted to know the facts. That the program was an abject failure, that the government wasted several millions of dollars for a unit pet program.

The executive agent had no clue. They were happy having passed their DoD responsibilities to the services and denied any responsibility in this adoption hysteria. Not once was the operational or strategic value of the program ever challenged. This was purely an opportunity for politicians looking for a human-interest story to bolster their images and manipulate the public for the almighty vote.

The Effort Continues

By mid-February, Jimmy and I were ordered to K2 and directed to remain until all TEDDs were moved out of K2 facilities. We had made strides with law enforcement week adopting forty-seven dogs to law enforcement and another sixty-three were selected for worldwide assignment to Army installations. Nine had been euthanized due to medical conditions. We had been able to adopt nine to former TEDD handlers and private citizens adopted thirty-one. We still had TEDDs to adopt at K2 and the forty-one that were supporting Brigades in Afghanistan.

One blessing is that Army Contracting Command did find some wiggle room to extend the contract to Mid-March 2014, a thirty-day reprieve. As you can imagine, K2 was not pleased, anticipating that the TEDD contract would be extended for another option year. This was a part of their profit margin and now the rug was pulled out from under them. Even the Marine Corps was miffed at how the Army could just shut down the contract and smoothly transition 229 dogs in less than two months. K2 recourse was to let the Army know that all Army owned dogs must be off their property on the final day of the contract, no exceptions.

We took a lot of heat from *Justice for TEDDS* and many of the former TEDD handlers stating that they were never given the opportunity to adopt "Their Dogs." First, TEDDs were not their personnel property, which is what they projected to the media and members of Congress. Second, we had a list of former handlers who we tried to contact. Many who we contacted expressed a desire to adopt a TEDD but were unable to. Why? Because they couldn't afford to pay for travel to K2 or they didn't have adequate homes for the dogs, or they didn't have funds to care for these dogs. That was not the Army being mean or disregarding these TEDD handlers. It was a matter of being empathetic, but we had a mission to do. Unfortunately, everything related to the adoption process through *Justice for TEDDs* painted me and Rob as villains. We were the top two most wanted persons in the country.

What no one realized is that adoptions were not a quick process. All dogs by regulation needed to be provided an adoption physical and neutered. Rob collaborated with the Vets at Ft. Bragg and transported TEDDs back and forth in his own car to make sure they were medically cleared. This included bite muzzle videos the vets needed to evaluate the dog's aggressiveness. And the neutering and recovery process took time. We were overwhelmed. To try and expedite the process, we brought Army Veterinarians from Camp Lejeune to Bragg. But even that was a Band-Aid fix. In the end, we completed adoptions for all dogs at K2. All "hold harmless" agreements (releasing the DoD from any liability) were completed. But completing the neutering procedures for all TEDDs could not be accomplished. Adoptees were presented with an Army Veterinary IOU for the new owners to take their dogs to a military installation and have the neutering procedures completed. The clock ran out.

In accordance with the Army's guidance, at approximately 1900 on the 15th of March, Jimmy and I pulled out of the K2 facility in a van loaded with dogs destined for Ft. Meyers, VA. We were transporting dogs that would be held temporarily until they could be transported to Korea.

We then faced a public relations campaign against the Army. Accused of total mismanagement, the poster child of TEDD adoptions became an Army specialist and his father pulling out all stops. The media and elected officials all rallied around, garnering national attention. The names will remain anonymous.

OPMG did receive a telephonic query from the father of a former TEDD handler inquiring on how his son could adopt the TEDD he severed with in Afghanistan. The father informed me that his son was eligible under Roby's Law to adopt his former TEDD. I took his information and investigated. What I discovered was that the TEDD was not eligible for adoption under Roby's Law. Roby's Law states that if a military member or their assigned MWD are wounded in action the dog can be adopted to the handler.

That was not the case for this TEDD. The specialists TEDD was not, I say again, not wounded in combat. The dog was suffering from a non-combat medical condition and was evacuated to the states for follow-on treatment. We also reviewed the casualty reports for the period when the specialists father stated his son had been wounded. There was no record of injuries sustained in combat. Further investigating showed that the soldier's injuries were not combat related. He did not meet the eligibility requirements.

But the father had gone to the news media, and members of Congress to take up this cause. The truth didn't matter. The air waves and social media attacked the heartless Army of prohibiting a wounded dog to be adopted by their combat wounded hero handler. We received daily inquiries from members of Congress and responded with the facts. Even today under HIPA I am not allowed to provide any identifying medical information relating to the specialist. The facts were clear, the dog and handler did not meet the adoption criteria under Roby's Law.

As this continued to capture media attention, we executed the adoption suitability checklist. What were the conditions of the home, his capability to take care of the animal? What we discovered was that he was disqualified from adopting a dog. One, because of the injuries he suffered, by his own admission, he was unable to provide proper care. Making him ineligible. Second, they did not have adequate facilities to maintain the dog. They had no backyard fence. With the intense media attention, the American Humane Society came to his aid and paid almost $10K to adequately prepare the yard to meet DoD adoption criteria.

Investigations

This situation continued to gain attention. The young man's father made allegations that he had contacted the Veterinary Clinic at Fort Bragg—and a young Specialist supposedly claimed that she had processed the TEDD's adoption paperwork and submitted it directly to me via email. This led to the first of two Army Criminal Investigative Command investigations against me. I was interviewed and while not charged, I was informed that CID was going to investigate these allegations and go through all my emails. This was a priority interest for the Army. Not only was I being tarnished on Facebook, but the Army was looking to pin the tail of fault on me. Every day was frustrating. And I was still knee deep trying to get TEDDS adopted and plan for what to do with the TEDDS deployed in Afghanistan.

I had done nothing wrong, and I stood by my statement. I had never received an adoption packet from Ft Bragg. Unfortunately, the young specialist was on the radar screen, being harassed on emails and telephone calls. This young specialist was thrown into the mix, and being unjustly chastised for an adoption that wasn't even justified under Roby's Law.

Even with the OPMG family displaying encouragement and supporting me, TEDD inquiries from members of Congress were continuing to flood the Pentagon. Regardless that even after the CID investigation exonerated me, I remained the most wanted person and poster boy for all bad things TEDD. I was the number one most hated person on the *Justice for TEDD*s website.

In 2015, the New York Post ran an expose on TEDDs making accusations of negligence, demanding those involved be punished. I didn't have Facebook at the time, but my colleagues shared Justice for TEDD comments from posts all around the world decrying me as a criminal. Because of immense political pressure, the civilian that had legally adopted the former TEDD caved in and agreed to give the dog to the Specialist. After three years of nonstop attacks and accusations, the Secretary of Defense directed the Office of Inspector General to conduct an independent investigation on the TEDD program.

Post K2

What no one cared about or realized was that after the media frenzy calmed down, we still had dogs waiting to be picked up. No one wanted to know how that magic was happening. Rob went above and beyond the call of duty to make sure the remaining TEDDs were united with their new families. He should be sainted for the extraordinary personal efforts he and his family made. Of course, there was no one to blame so no one gave a shit about it. Only the pandering media and inquiries from the hypocrites from Congress gained the attention of Army leaders. Here is the back story of a real canine hero.

Rob was a retired Non-Commissioned Officer. He was of the troops for the troops. He absolutely adored his soldiers. In a totally unauthorized and unorthodox move, Rob temporarily took all TEDDs that were pending adoption to former TEDD handlers and civilians and brought them to his house! Yes, his house became the TEDD kennel. At his own expense, he procured dog food, continued to take dogs to Ft. Bragg for adoption physicals and to get neutered. That meant his family was also part of this extraordinary effort. Keeping dogs for up to two weeks on his own dime so that former handlers would be able to decide when and if they could pick up "their dogs."

But Rob was already convicted in the court of public opinion, indicted as my accomplice. For his zeal and dedication, he was trashed by *Justice for TEDDs*, dragged through the media circus, accused of giving dogs to friends, his house stalked. Then Murphy reappeared and Rob's contract at FORSCOM was not renewed. But did Rob turn his back…no, troops and dogs still needed to be taken care of.

Remember the forty-one dog teams we deployed in Afghanistan? Did anyone from the Media, Congress, or *Justice for TEDDs* demonstrate any interest in how those dogs would be adopted? Nothing. It didn't make a good news story. But miracles do happen, welcome the TEDD Santa Claus.

Problem #1. While the dogs were deployed, they remained Army property. But the normal scenario was that once the dogs completed their service in Afghanistan, they would be returned to Charlotte International Airport North Carolina. There a contract representative from K2 picked them up and returned them to their training facility in Southern Pines, NC, to begin training with the next unit. Where would the dogs go now? There was no contractor. That didn't change the pre-arranged flight returning the dogs to Charlotte.

Problem #2. The dogs were returned to Charlotte, but their handlers were not. Handlers passed their dogs off in Afghanistan and then returned with their unit to their home base. Remember that once the dogs landed in Charlotte, they were no longer Army property. Who was going to pick these dogs up and provide care for them when they arrived in Charlotte?

The solution–execute the adoption process in Afghanistan. We had to employ theater kennel masters and veterinarians to perform physicals and complete adoption documents in Afghanistan. Here's where soldiers shine. SFC Lavalle, now the first female Army MWD Sergeant Major, was the MWD Program Manager at Bagram Air Base, she took the lead. With SFC Lavalle initiating the paperwork and working with the vets to execute the physicals, dogs were prepared to rotate back and be immediately adopted. We had several months to affect this plan. In all cases, TEDD handlers who elected to adopt their assigned TEDDs were accommodated. We also had TEDDs designated for the U.S. Capitol Police, and we completed their adoption packages. The question: if there was no contract, who would take care of these dogs until handlers and law enforcement agencies picked up their newly adopted dogs? Who would provide temporary shelter and feed them?

Saint Rob became the Angel of the TEDD program, but no one wanted to hear about the personal sacrifice he was making to ensure that dogs were reunited with former handlers. No one wanted to acknowledge his personal expenses. He became the TEDD Santa Claus. As a private citizen, he picked up every returning TEDD at Charlotte. If a former TEDD handler was unable to pick up a dog due to duty requirements or lack of funds, Rob kept the dog at his house or put the dog in his truck and drove thousands of miles to reunite the dog with their new owner. But that wasn't a story that made headlines. God forbid the media tell this story of the Army Canine Santa. Yes, was I aware of what was going on and that Rob was doing this on his own. Yes, the mission had to be done, and Rob and his family were the chosen ones. His reward–continued attacks by the media.

Post Contract –Accusations

Yet the Executive Agent remained like a Teflon frying pan. Any responsibility for failing to meet their charter, slid off the fifth floor of the Pentagon, and landed squarely on the shoulders of the Army and Marine Corps. The EA was never brought to task for relinquishing their DoD program oversight. The Air Force maintained a safe distance while the Army faced daily scrutiny. The EA's culpability in intentionally failing to provide sufficient dogs to meet the services demands.

Let's clearly articulate what happens when liability is transferred from the government to an agency, former handler, or private citizen. The culmination is the signature by the adoptee on a 'hold harmless agreement.' The adoptee executes a legal document, effectively removing all liability from the government. Once executed, the government is free and clear. Once liability is transferred, no further action by the DoD is required. There are no post adoption follow-on inspections to perform health and welfare checks. The process is clearly articulated.

For Law enforcement agencies, the process was to provide a letter from their department stating their need for a retiring military working dog. These letters followed a standard format, letterhead stationery, signed by their department chief or designated representative. The letter outlined that veterinary care would be provided, and adequate shelter provided. Based on those criteria, law enforcement kennel masters or trainers screened dogs, and we would execute the hold harmless" agreement. Transition completed. However, in the eyes of the media TEDDs were always going to be an Army responsibility.

The Taylortown North Carolina Police Department had presented an official letter signed by their Chief requesting excess military working dogs. All paperwork was reviewed and in order. Following the process, they selected six dogs for adoption. We executed the hold harmless agreement on 7 February 2014. We performed our due diligence. That should have been it, but it wasn't. Shortly after adoption, there were allegations by residents that the police department had been mistreating the dogs. One of the media outlets picked up the story and it was another field day on the Army. Inquiries from the house wanting to know how these dogs could have been adopted were levied against OPMG. Legally, the issue was no longer a military matter. Post adoption was a civilian animal control issue. However, we were on the chopping block. The Army should have had some crystal ball to project the long-term treatment of the dogs! The cloud of public opinion hovered over the Army as if it was an acceptable conclusion that anything TEDD related must be the Army's fault.

One decision that I regret was the adoption of thirteen TEDDs to Soliden Technologies. Soliden presented themselves during law enforcement week, with a letter requesting TEDDs to train as service dogs for veterans. They did not meet the adoption criteria for law enforcement. I believed that their intentions were honorable, but legally I was not sure if they fell into any category to adopt a TEDD. So, I reached out to the Executive Agent, presenting my concerns and requesting a legal opine. The response I received was that the dogs did not meet the adoption criteria for law enforcement. However, they could meet the private adoption criteria. At the time, there were no more law enforcement agencies requesting dogs or former TEDD handlers, either on the list or who had contacted OPMG. Their paperwork was in order and with that legal opine, I was confident these adoptions were legally sound.

This decision came back to haunt me, a decision I made based on documentation and the legal opinion received. I did my due diligence. I believed I was ensuring the safety, health and welfare of the dogs for the purpose of supporting our veterans.

Soliden Travesty

Approximately a year after the adoptions, a private kennel in Virginia contacted me, wanting to know how he could obtain funding from the Army to pay for thirteen former Army dogs, left at his kennel by Soliden. This was a total shock, I couldn't believe it, what was this all about?

The owner informed us that he'd been contracted to kennel the former TEDDs on behalf of Soliden. He informed me that Soliden had requested him to kennel the dogs until they could arrange how they were going to use them for their "Business." Business? The hold harmless agreement clearly states that adopted MWDs will not be utilized for any post-retirement business use. Their "hold harmless agreement" technically was now null and void. This kind gentleman had been taking care of these dogs, incurring an expense of almost $50K. We had no ability to provide any funding. These dogs were not government property.

This lit the fire and the news media and members of Congress jumped all over it. The Army had intentionally allowed this to happen! This claim restarted charges of the Army mismanagement, denying former TEDD handlers the ability to adopt the dog they'd served with.

What no one knows is that while this media frenzy was attacking the Army, I'd already set in motion a course of action to get these dogs adopted by former handlers. I already contacted an agency that I knew would be able to correct this nightmare. *U.S. War Dogs of America*, whose owner and Executive Director, Ron Aiello became a savior for these dogs. Somehow, Ron provided free medical treatment and prescription medications for retired military working dogs across the nation.

I reached out to Ron and informed him of the situation asking if he could help to re-patriate these dogs. I agreed to provide him with the names of the former handlers. Ron jumped on this and set it in motion. *War Dogs of America* made it happen, but at a cost. All former handlers, minus one were reunited. Ron had to negotiate with the kennel, who would not release the dogs until he had received $50K in compensation to pay for his services. Understandable, but we found out that Soliden had dissolved, and their board was nowhere to be found.

But Ron, a former Marine Vietnam Scout Dog handler, wasn't going to let these dogs linger. He worked out a deal and paid the kennel $20K from *War Dogs of America*. The Army...through the compassion of Rob Squires and others...took care of soldiers and dogs. The public only saw one side of the TEDD program. During this short fuse process, we attempted to meet our orders under very difficult circumstances. However, the record books only record the shortcomings of adoptions, cultivating incompetence and placing blame.

What this should have done was to open the eyes of the SECDEF that it was AF EA failure that led to this debacle. This should solidify that "part" time handlers are an exercise in futility. Partially trained detection teams are dangerous and ludicrous. Millions were wasted for a program providing zero bang for the buck. A plan to provide for surge requirements and the manpower and resources is what is needed by the EA. We train the best of the best handlers, great troops. We need professional MWD teams, not an ad hoc, off the wall program, that in the long run will place lives in danger.

Post Contract Monetary Claims

Regardless of the contract being concluded for almost two years, discrepancies over finances continued to haunt the Army. The prime contractor, DPMS, failed to pay K2 Solutions for services at the conclusion of the contract. This figure was well over $1M. K2 approached the Army requesting our intervention to attain payment. The Army was not the financial manager of government funds and had no legal authority to intervene.

This was a contractual issue between DPMS and K2. But this issue again became my Sword of Damocles. K2 made allegations that the Army had promised additional funds to expand their kennels. The allegations focused on a meeting where supposedly I had agreed to this. There was no written agreement or meeting minutes that provided any evidence I had agreed to allocations of government funds. OPMG legal concurred and opined that the allegations were baseless. Case closed? No. Apparently through a review of the contract by the Army Contracting Command, it was determined that the contract language, as an OPEN Contract, allowed the contractor to bill the Army for any directly related expenses. Of course, it was determined that the billing for the additional $300K regardless of fault or lack thereof, that OPMG was liable, forced to pay K2 the additional $300K. TEDD would not go away.

Inspector General Investigation

Given the overwhelming social media engagements and concerns of improprieties with the adoption process, Congress directed the DoD to perform an investigation into the TEDD program. In 2016, the DoD Office of Inspector General commenced their investigation. The postmortem was not an investigation of the failure of the EA to provide dogs to meet warfighters' demands. It was centered on Army procedures—in particular, the adoption process. It never addressed the underlying root cause of who was responsible to procure and train dogs for the services. The EA dodged the bullet of responsibility while the Army Office of Provost Marshal General bore the brunt of the blame. The report was released on 1 Mar 2018. The conclusions (summarized):

1. The Army failed to develop a transition plan.
2. The Army failed to prioritize adoptions, allowing non-traditional handlers priority for adoption.
3. TEDDs were not accounted for in the DoD Military Working Dog Management System.
4. The Air Force failed to provide proper oversight for non-traditional dogs.
5. Army and Air Force policies should be revised to include adoption requirements for non-traditional dogs, and non-traditional dogs should be accounted for in the WDMS.

Holistically, this should have spawned further investigation of the entire DoD canine program. But that did not happen. The EA remains untouchable, while the trained-dog deficit continues to grow, impacting canine readiness to this day.

Not once was the Air Force chastised for relinquishing their responsibility to procure and train dogs for the services. The IG report directed the Air Force and services to modify their guidance for the inclusion of a non-traditional handler adoption process. It also included the requirement to capture these dogs in the DoD tracking Data Base, the Working Dog Management System (WDMS). All useless administrative policy corrections.

The IG report also resulted in the 2017 National Defense Authorization Act which directed the services to pay to transport adopted dogs…returning from overseas locations…to the U.S. However, the legislation failed to provide additional appropriations so the services could pay the approximately $10K it took to return a dog to the U.S. The services MWD programs now had to determine which area of the program they would pull the money from to pay for returning adopted dogs. Do we cut back on dog food? Training aids? Which pot do we rob from Peter to pay Paul?

While the OIG report recommendations have been implemented and policies updated to reflect adoption of non-traditional programs, the Air Force remains unable to meet its traditional obligation to train dogs needed to support the DoD program. They failed to meet the trained dog requirement for a decade (2013 to present). And the gap continues to grow. We can only hope that…should we be involved in another major contingency mission…the Executive Agent will not be caught napping again, ignorant of the strategic needs of the services to support global contingencies.

Retribution

Throughout the history of our nation, brave military leaders pushed for a new vision. For example, in 1929, General Billy Mitchel challenged the "old school military" and threw away his career to fight for Air Power. His vision and passion eventually led to the creation of today's Air Force. General Patton espoused his concerns as World War II was coming to an end, that the true threat to the free world was not the defeat of Nazi Germany, but the threat of Communism. The blame for this dysfunctional failed program was no different. Place the blame on the lowest common denominator. While Generals pounded their chests and threw millions on a totally failed program, when it ended and went sideways, they pulled on their Teflon overcoats and hid in the shadows. Not one senior leader—especially the Air Force Executive Agent was on the blame line. I was Billy Mitchel and George Patton of the TEDD program.

Along with Rob Squires, I was the bureaucracy's sacrificial lamb. Our names are forever etched as TEDD villains. Proud of being the bearer of the truth, I brought my experience as the only living person to have run canine combat operations in a theater of war to the DoD, and for my experience and attempted leadership I joined that club of visionaries, that was quick to pursue their attacks, codifying me as having gone "native."

A small retribution was the fact that Von Liche Kennel (VLK) was charged with fraud. On August 14, 2020, the Northern District of Indiana accepted settlement from VLK in the amount of $1,350,000. The claim validated that VLK failed to provide adequate levels of training, materials, and supervision for the Army's TEDD program. They also found that the Field Service Representatives (FSR), co-located in Afghanistan, failed to provide adequate oversight in monitoring TEDDs to reinforce and improve their operational skills. The FSRs should have been charged with negligence, as their jobs were to evaluate performance, but they never went outside the wire. However, there was no public outrage on the main root causes of TEDD failure. It didn't sell papers. It wasn't a human-interest story.

I stood up for what I knew to be the truth and called out the program's faults. Every claim I made of contract fraud at CENTCOM and to the Army was accurate. The complaints, standing up to senior leaders and calling out VLK were all but forgotten. The public wanted someone to pin blame for TEDD…and Rob Squires and I were it. Everything that was clearly articulated leading to the failure of the TEDD program was codified. VLK was held accountable. It was not due to the testimony of those directly involved in the TEDD fiasco. It was from the internal testimony of former VLK employees whose conscience would not let injustices go unpunished.

TEDD Epitaph

Non-traditional programs are not effective, period. The Marine Corps program functioned better than TEDD on a limited basis but faced the same shortfalls as the TEDD program. What should be in place is the EA and COCOM have a strategic surge plan and appropriations to provide additional MWD teams. We failed to learn from the past. The dogs for defense program, the WWII program provided thousands of dogs to meet the war effort. That program end justified the means and provided a surge requirement to support two theaters of war and domestic force protection. TEDD was a quick fix and assumptions by senior leaders of a hocus pocus wave the magical wand fix. If the DoD is not held accountable to execute their DoD directed mission, there will be another "Shit Show" with lives in the balance.

When the balloon goes up…and it will…the United States Military will need an influx of dogs. They will need to build a force structure, possibly of counter-suicide, medical mitigation animals. But the Air Force will remain unprepared. Like the ostrich with their head in the sand, they'll throw up their hands in surrender, incurring mini-EAs passing the torch to the services. And people will pay with their lives.

Mandated Missions and Operational Support

Chapter 14 – U.S. Secret Service Missions

By law – "The Department of Defense will provide services, equipment, and facilities in support of the US Secret Service (USSS) when performing its protective duties under Section 3056 of Title 18, United States Code (U.S.C.), and Public Law 94-524 and as directed by the Secretary of Defense in accordance with the requirements of Department of Defense Directive 3025-13." In other words, the DoD provides assets for the President and Vice President for all official travel and identified National Security Events.

These missions are divided into two categories: Standard Dignitary Support and National Security Special Events (NSSE). Both support the USSS with the only distinction being that standard missions are paid out of the military service travel funds and NSSEs are reimbursable.

This means that under the law, military working dogs are one of the critical elements supporting USSS domestic and global protection missions for the President, Vice President, and State Department - Secretary of State. Dignitary support encapsulates approximately thirty-five percent of an MWD teams' duties and responsibilities. This along with their Title 10 United States Code [Reference 11] missions which provide detection teams in support of:

1. Overseas contingency operations
2. Special missions in support to federal law enforcement,

The public has no idea of the scope of these missions. We place the safety and security of our national leadership in the hands of young men and women, some only nineteen years of age. We trust them to mitigate possible explosive threats. Canines use their olfaction to detect explosive components. And handlers are trained to read their dogs' behavioral changes. They are a two- and four-legged detection system.

The orchestration of a USSS support mission starts at the USSS Headquarters Technical Support Division in Washington DC. The "TSD". They are the hub that directs all dignitary protection requests. They come in three forms:

1. The White House USSS Detail requests DoD resources to provide protection for POTUS/VPOTUS travel dignitary protection.
2. The White House Military Office, commonly referred to as WAMO would request travel support from TSD to provide security for Air Force One.
3. Secretary of State Office of Dignitary Protection would request military resources through the Air Force Tasking cell.

For the most part, these requests provide supporting military agencies with a few days to ramp up teams, make travel arrangements, attain visas, Travel Health Certificates for MWD's (good for ten days of overseas travel) and verify any special country travel issues, such as quarantines, or special medical tests. The coordination should flow smoothly. However, that has never been the case due to the inability of the Air Force…as the Executive Agent…to change fifteen-year-old protocols to protect manpower and retain their prize possession at the Air Force Security Forces Center (AFSC) Center at Lackland AFB, Texas.

There are two elements of the military that comprise dignitary support: MWD teams and Explosive Ordinance Detachments. The Explosive Ordinance personnel are assigned to the Washington based,

"Very Important Person Support Activity" or VIPSA. MWD teams are located at military installations across the globe. Both have the same mission:

1. Clear and detect explosive threats at venues.
2. Be prepared to disarm and render safe explosive threats (Explosive Ordinance Disposal teams).

Tasking Cell

The striking failure is that the Air Force ignored continuity of operations, retaining the USSS Tasking cell under the AFSF Center at Lackland two thousand miles from the mission hub of the USSS TSD. Continuity and integration with the mission command center is maintained via emails and telephone conversations. The Tasking cell was attempting to coordinate missions two thousand miles away, while their VIPSA EOD counterparts have liaison teams in TSD headquarters..

Why is the Air Force so determined to retain control of their tasking cell in Lackland? Simply stated, it's the "Prize." And along with the prize are manpower positions that could affect the Center's civilian and military manpower authorizations. Add to the equation, I was the DoD Single Point for all things Canine to the Executive Agent, and as I pointed out, my job description clearly articulated that USSS dignitary protection mission was under my "sole" responsibility. But AF cronies told me that Secret Service dignitary protection was *not* my job. The subject matter expert for the DoD is not in the MWD chain of command. Non-canine civilians and full Colonels with zero understanding of the administrative and logistics requirements to execute a Secret Service mission are. The trend of non-support of the MWD program was becoming more and more clear. It was about the best interests of the Air Force.

The explosive ordinance cell is task organized to be immediately responsive, located in DC, self-sufficient, able to perform administrative tasks, such as attaining VISAs, processing travel orders, and generating travel reimbursements. While their efforts are transparent under a centralized command team, the AF tasking cell has no such capability, each mission is a juggling act. The tasking cell has no command authority to direct the services to source teams. Their location in Texas prohibits them from attaining VISAs/travel documents directly from Embassies in D.C. The coordination with D.C. Embassies is tasked to the respective military services who prepares VISA applications and mails the VISA application and Handler Official passport, rolling the dice that the respective embassy will have sufficient time to turnaround the VISA and mail it back to the unit before the date of the mission.

For National Security Special Events, VIPSA (EOD) centrally processes all reimbursable travel vouchers and works directly with the USSS to collaborate projections for annual NSSE funding. The AFSF Center and Tasking Cell do not coordinate funding projections, guess who has that responsibility… the DoD MWD Program Manager. The Tasking Cell—as is the 341st Training Readiness Squadron—is mis-aligned. Both support a joint service mission but is not staffed or equally represented with Joint Services oversight and staffing. The basic principles of military operations—unity of effort, continuity of operations, economy of force…and most importantly…their ability to understand Joint operations—is intentionally stonewalled by the Air force.

In June 2013, the JSMWDC conducted its semi-annual meeting at Lackland AFB. We met in the conference room of the Security Forces Center. A discussion and recommendation to relocate the Tasking Cell to Washington DC… for continuity of operations and to eliminate administrative and logistics gaps …was on top of our agenda. A full Colonel from HQ Air Force, representing the Director of the AFSF, listened intently to the Joint Services concerns. At the time, I was assigned to the Army as their MWD Program Manager and was a huge proponent of re-missioning the Tasking Cell. We were told that this recommendation would happen. We were promised it was a done deal. As of my retirement as the DoD MWD PM on 1 Apr 2020, that action remains open. Transparency remains at arm's length. The Air Force continues their tap dance. Three fine Air Force NCOs are held accountable for the success or failure to protect our Top Three National Leaders.

VIPSA (Very Important Person Support Activity) is functional, and task organized to meet their mission. Located in Washington D.C., this unit…consisting between eight to twelve joint service explosive ordinance technicians…is dedicated to a standard two- or three-year permanent assignment. These joint service specialists receive special VIPSA credentials and deploy to their assigned missions to and from Washington. VIPSA has an operations team co-located with the Secret Service, who immediately coordinates and assigns assets. Strategically, being co-located gives them a heads up on current and projected missions providing a buffer to perform long range resource planning (USSS projects missions on quarterly cycle).

Administratively, as opposed to the AFSF cell 2000 miles away, VIPA admin personnel frequently hand carry VISA applications to D.C. embassies, expediting the VISA process for overseas travel. VIPSA maintains a fantastic working relationship with the embassy staff. The AFSF Tasking Cell, two thousand miles away, has no interface with the embassies. Embassy coordination is placed on the shoulders of the services kennels across the nation and overseas to independently process their own VISA applications through the postal service. Zero continuity! Of course, nothing can go wrong with the postal service.

Continuity doesn't seem to be part of the AFSF language. In contrast, with VIPSA who has two liaisons assigned to the USSS TSD in D.C., the MWD desk sits empty, collecting dust as it has for more than ten years. Because of this close coordination and being co-located in the DC area, explosive technicians can quickly grab their "Go Bag" and with the support from their admin team, immediately deploy. MWD teams are at a huge disadvantage. The explosive technicians can travel with the Secret Service on their designated "Car Plan" (support personnel/equipment) out of Andrews AFB, while our canine teams must make reservations for commercial travel, rental cars and hotels. Independently travelling with their dogs and equipment. We did have an opportunity for MWD teams in the D.C. area to fly on the car plane but was always on a case-by-case basis. We were still stragglers.

Who has oversight of the tasking cell? A GS 12 employee of the Security Forces Center. Not an MWD handler or someone who understands MWD operations. The employee has never met or engaged with the USSS Technical Support Division at their headquarters in D.C. The employee is not available to the Tasking Cell 24/7, because he/she works a normal eight-hour day and is unavailable on weekends. This misalignment and inability to maintain a 24/7 chain of command is ludicrous…and not supportive of the Joint Services. Yet as the DoD PM, this was part of my job description. I was on call and responding 24/7, ensuring mission accomplishment, as were the service program managers. Yet as a GS-14, I was not within the realm of AF royalty to have a government cell phone to monitor and direct worldwide missions. The AF operated in a reverse mode, position not the mission was the priority.

What can and has gone wrong in the last five years? First the Tasking Cell is not staffed with MWD handlers. The assigned personnel are AFSF personnel—three sergeants. The manning document authorizes ONE. The other two are "additional duty positions". There is no established Standard Operating Procedure outlining duties and responsibilities. These NCOs are just expected to "learn on the Job." The Operational Tempo is intense…over three hundred fifty missions annually, and during an election year an additional one hundred missions. A newly assigned NCO is basically thrown into the mix and either sinks or swims. However, should an NCO fail to adapt to the high-profile mission or not grasp the multi-functional responsibilities, it will be reflected in his performance report. Or the airman will be relieved, and another un-trained airman will be thrown to the wolves. By the grace of God and because of the dedication and resilience of these fantastic NCOs we never experienced mission failure.

These NCOs were told by AFSF leadership NOT to reach out to me as I was not in their chain of command. Ignorance is bliss and I continued to direct the tasking cell. Along with the MWD Program Managers, and our joint MWD teams, they continue to maintain the mantra of "Failure is Not an Option." It was these young men and women who independently displayed exceptional leadership, sound judgement and discipline in a multi-agency global environment and their missions were accomplished. While AF leaders used them as pawns.

National Security Special Events – MWD Liaisons

For all National Security Special Events (NSSE), the Secret Service requires the military to provide a Liaison NCO. This is different from a "standard mission" NSSEs, which you'll see in the chart below, encompass…in many cases…more than one hundred dogs. Provided through a fair share breakout by each of the services. The biggest challenge for the MWD assigned liaisons is that this mission set is not taught in any MWD handlers, trainers, kennel masters, or advanced MWD courses. My understanding is that the services are starting to include this in MWD advanced training, but at least until 2020 our Liaison NCOs learned this high visibility mission on their own.

As a Secret Service Liaison NCO, they are immediately thrust into a demanding situation. For many, this is the first time that they are in "command." The learning curve is immediate with no room for error. The LNO has the overall responsibility for the health welfare, accountability, mission support and discipline of all the assigned MWD teams (Handler and Dog). They must adapt and roll with the punches. It doesn't matter if they've never done this before. The Secret Service expects missions run like clockwork. The breakout below describes the 2019 MWD taskings.

1. Republican National Convention - 135
2. Democratic National Convention – 135
3. United Nations General Assembly – 75
4. G7 Summit – To be Determined.
5. Inauguration – To be Determined (2017 – 120)

Yet our young NCO's worked magic. Many arrived several days prior to the event. Without a site visit and with minimal logistics or administrative coordination, they walked in cold, unaware of travel arrangements, billeting (hotels), vet support, communications, and credentials. The burden of responsibility rests with these great NCOs who give 150 percent. They do what NCOs do—adapt and overcome. Working arduous hours without time to breathe. Without fail… they got the job done.

Another significant issue contributed to last-minute coordination problems. There was a non-existent relationship between the Tasking Cell Supervisor and the Secret Service Technical Support Division. For years the JSMWDC invited the USSS TSD to attend our semi-annual meeting, and for years, they never attended. The Tasking Cell supervisory chain refused to leave the comforts of Lackland AFB and establish a positive relationship. The atmosphere between the Security Forces Center and the Secret Service was strained.

That changed in 2019. We established a positive relationship and opened the doors with our brethren at the Secret Service Technical Support Division. We dramatically changed paradigms from negativity and mistrust to a unified partnership. I established a standard monthly meeting with USSS TSD. We needed to rebuild. We needed to build trust and remind each other that we were cops. I started a new concept, monthly meeting with the TSD Special Agent in Charge and coordination planners. If nothing more, to have coffee and chat. Friendships were established, favors asked, information shared openly. From that day forward, we were no longer adversaries. We were partners.

We concurred that planning and NSSE preparation could not happen two days prior to the arrival of MWD teams. We agreed that pre-event site visits with lead agents and MWD LNOs needed to be accomplished at least three months prior to the event. We agreed that mission analysis and information sharing would have to be ongoing, and that the selection of LNO and lines of communication should be established immediately upon NSSE notification.

We exercised this concept at the United Nations General Assembly in September 2019. Two stellar individuals…the Air Force MWD Program Manager and a Canine Sergeant from our recently gained DoD civilian police agency; the Pentagon Force Protection Agency…were selected as the LNOs. In what is now part of the NSSE standard operating procedures, the LNOs were brought to DC and introduced to their TSD counterparts and the lead agents from the NY field office. Four months prior to UNGA. Plans were made to conduct site visits to New York City. The LNO's sat across the table and laid out initial MWD requirements. A critical lesson learned. We were walking alongside, rather than trying to catch up to our USSS partners.

USSS UNGA after action reports lauded this change in focus, exalted and praised the LNOs for their superior job performance. This reversed DoD trends of negativity and USSS mistrust. This also allowed the services to address the LNO training issue. We moved forward with a LNO "train the trainer mentorship program" that the Army introduced at UNGA 2017. Services would select a "mentee" to mirror one of the LNO's. They would "learn the ropes" providing another qualified LNO to the pool of NSSE LNO candidates.

These initiatives broke barriers. And then, in October 2019, the JSMWDC welcomed a new stakeholder…the U.S. Secret Service! It was the first time the USSS participated in a JSMWDC meeting. Their lead agents for the Republican and Democratic National Conventions provided a mission analysis, and we commenced planning, immediately selecting which service would provide the Liaison NCO. Side by side with the USSS leads, we set the stage for success. Transparency rather than last minute discourse, initiated eight months prior to the event.

Chapter 15 – Defense Support to Civil Authorities

Commonly known as DSCA, the military—through interagency agreements and memorandums of understanding with the Secretary of Defense—provides military personnel (units) and equipment to support the civilian community.

We're on the precipice of a new dawn in using military assets to support law enforcement in a more exclusive role. This new day will require major revisions to the authorities to employ military assets. The current regulatory authority for employing the military is taken directly from the Department of Defense Directive 3025.12 [Reference 12]:

Military personnel in Federal service pursuant to Title 10, United States Code (U.S.C.) (Reference (g)), will be under the command of...and directly responsible to...their military superiors and will not be used to participate in civilian law enforcement activities, including arrest, search, and seizure, in violation of section 1385 of Title 18, U.S.C., also known and referred to in this volume as "the Posse Comitatus Act," as amended (Reference (h)), unless otherwise authorized by law. During domestic operations, including DSCA, military members must be aware of the laws, Executive orders, and DoD policies concerning the handling of intelligence information.

That's changed since the attack on the Capitol building on January 6, 2021. The National Guard, normally mobilized to support their home states, was mobilized from five different states...and deployed to protect the Capital. In the United States, the Guard was mobilized and armed. Approximately twenty-five thousand National Guard personnel provided "Armed Security" for our nation's Capital. How were they mobilized? Were they mobilized under the governor's state authority under Title 32 USC [Reference 13]? Or were they federalized by the Secretary of Defense under Title 10? What were their rules of engagement for the use of deadly physical force? Under whose command were they being employed...the military, or federal law enforcement?

And that was not the only time in the past five years that we saw military personnel mobilized to support the civilian population. The National Guard, Reserve and Active-Duty Personnel were mobilized to fight the pandemic, augmenting the overwhelmed civilian medical community. These mobilizations were reactionary, but necessary to maintain national health and the security of our nation. The takeaway—there were no inclusive plans to use military manpower or equipment. Status quo, self-sufficiency was and is our Achilles heel.

Even now, interoperability, joint military – civilian training exercises are limited to how the National Guard can be integrated at their direction in a state emergency. But not with the active-duty military, those plans remain non-existent. Should there be an inter-agency task force on multiple levels to develop civil-military integration? Yes. The mobilization of troops to secure the Capitol was reactionary. The amount of force protection, personnel and equipment were not contained in any pre-determined plan. The nature of the entry into the Capitol was neither planned nor expected. But should it have been? Should there have been a worst-case scenario plan? Should the Sergeant at Arms and Capitol Police have been planning for multiple scenarios and inter-agency support...yes. But complacency, failure to react to intelligence, lack of inter-agency collaborations...all contributed to the failure to protect the Capitol. Plenty of blame to go around.

Most blatant was the institutional belief that this could never happen. It was the same tunnel vision that led to the attack on Pearl Harbor and 9/11. Like Pearl Harbor, our elected officials believed that the U.S. Fleet was invincible, the thought of the U.S. being attacked in 1993 and on 9/11 was incomprehensible. Yet the intelligence was there. Everyone knew it was more than a possibility that these attacks were imminent. Yet senior leaders assumed that their positional authority in the seats of power gave them more wisdom and understanding of the geo-political threats than the experts in the intelligence community. The attack on the Capitol was on the same methodology, tunnel vision, and the same refusal to act on sound intelligence. There is no doubt that, Post 9/11, anyone believing that we will not require interagency support, officially or unofficially, is living in another dimension. The threat is too extensive for the U.S. to operate with blinders on. MWDs were just a mere drop in the bucket of the current force structure identified to protect the capital. But what is the number one priority threat to the Capital campus? Improvised Explosive Devices!

Do the Capitol Police have sufficient bomb dogs to provide force protection to the interior of the Capitol and the expansive Capitol campus? Will they be able to provide the standoff distance and exclusionary zones? The post January 6th report on the protests answered that question…NO! The threat of explosive devices will not be posited in areas highly patrolled. Who will serve as the Canine Coordinator? Who will evaluate patrol areas and assess secondary and tertiary searches? How will dogs be utilized…and how will MWDs be integrated into a formal plan, actionable, under a stable chain of command?

This is not unique to the Capitol but to every municipality in the nation. Gathering the plethora of agencies to develop a mutual aid agreement? Has there been any cross-agency training to develop relationships, understand detection capabilities, tactics techniques, procedures? How many law enforcement, military, and civilian dogs are in the region capable of responding? The answer is no clue. Offices of Emergency Management do not maintain centralized data. "We'll figure that out after the incident occurs" is not the answer. This needs to be achieved through a shared national database. A flip the coin methodology to identify availability is not a viable solution. Interoperability, not a a knee jerk reaction needs to be the mantra of all agencies. Like the Boy Scouts motto…Be Prepared.

PLANNING, PLANNING, PLANNING!

Take note, this scenario being played out at the Capitol is the same scenario that could possibly happen in major cities throughout this nation, individually or simultaneously. Are we prepared to meet these impending threats? No. And the reason is so simple…ego. No one wants to admit that they can't handle their own agencies 'security issues. We dispelled those attitudes very quickly in CENTCOM. Lives depended on teamwork. The adage, "One Team One Fight," remains the voice of unity. Government agencies, and municipalities continue to dispel the need to work together, and plan based on what ifs. Institutional beliefs that "our institution" can mitigate any threat or natural disaster without outside assistance are antiquated and unrealistic. Power plays result in lives lost.

If it can be thought of, our adversaries probably already have an execution plan in place. Regardless of how outlandish or outrageous, it is part of someone's thought process and needs to be addressed. Our adversaries don't play by the conventional play book. They operate asymmetrically (outside the box). Institutional thinking across the spectrum needs to change. The new normal is coalition partnerships and inter-agency collaborations. They can be the difference between life and death. My efforts to bring about this open-minded institutional thinking were met with the accusations of "Mission Creep!"

The tide has turned and the oath we take is clear, "to defend and protect the United States from all enemies foreign and DOMESTIC." Our canine community is miniscule in comparison to the DoD, but irreplaceable. Because of their limited numbers, these animals must be trained to one standard. Why… interoperable, trained to be ready to deploy to any location and go right to work. If we don't work to canine standardization, we will be a multiplicity of dog teams, chasing our tails unable to be cross leveled anywhere in the U.S. to support our first responder brethren.

The National Guard – One Army Team

The new world order has arrived, and Domestic Terrorism is out from the shadows. Meeting these domestic challenges rests with the states organized militias. The National Guard. Organized under Title 32 of the United States Code. They fall under the Command and Control of the Governors. Under this title they are empowered to enforce the laws of the State. Each state has a Guard force that supports the state during natural emergencies and acts of civil unrest. But there is only one state that has federal military dogs in their inventory. The Connecticut Army National Guard.

A unique concept, and the Connecticut Guard holds the distinct honor of having the only canine unit in the National Guard. An experiment in Army canine diversity that has proven immensely effective in supporting the state's civil defense and law enforcement missions. Their dogs have been fully integrated in the Army mission and the 928th MWD Detachment (Small) have supported numerous Secret Service Presidential and Vice-Presidential dignitary protection missions. and National Security Events.

Located approximately forty-five minutes from New York City they have been instrumental supporting POTUS, VPOTUS, SECSTATE and the UN General Assembly and the 2016 Presidential election. They have been the go-to unit for MWDs when POTUS, et al makes last minute travel arrangements to New York City. The 928th has provided direct support for the U.S. Military Academy performing random drug sweeps and explosive detection dogs to sweep Michie Stadium prior to Army football games. The 928th has also supported law enforcement, deploying explosives dogs to the Boston Marathon. They have even served in the naval service supporting the U.S. Coast Guard, providing narcotics dogs for counter drug operations on Long Island Sound ferries.

Under Title 32, the Governor can deploy the 928th at his discretion. The threat barometer has shifted, the primary threat to this country is now considered domestic terrorism. As such the Guard will operationally be called upon as the states force in readiness. Will they have all the capabilities needed to perform that mission? Is it plausible for the guard to look at increasing MWD detachments to meet the domestic challenge. A question that this author believes needs to be raised.

Parity

Should the military working dog teams be granted law enforcement authority when supporting a DSCA mission? This is a question that has been discussed openly within the military police community for a long time. Let's look. Military Police are trained to meet the mandates of the Civilian Police Officer training Standards (POST). In many states military police training is accepted as meeting the benchmarks of police officer recruit training. MPs in those selected states still must complete training in state uses of force, weapons qualification, and specific penal law codes. Police recruit training, in some cases, has been reduced from six months to several weeks.

When our dogs deploy to support law enforcement, we are given a search area. Under *Posse Comitatus*, deployed MWD teams are prohibited from active law enforcement. Any finds (explosive/narcotics) must be confirmed by a law enforcement agency with their dogs. If we engage a suspect, we are allowed to detain. Law enforcement must execute an apprehension, determine violation of law, mirandize and arrest. Is *Posse Comitatus* outdated? Is it time for *Posse Comitatus* to be re-assessed? *Posse Comitatus* became law in 1898 restricting military personnel from performing civil law enforcement. As we face a new horizon of "domestic counterterrorism" is *Posse Comitatus* still relevant? With the explosion of crime, a climate of "defund the police", could MPs be introduced to protect and serve the civilian community. A discussion is warranted.

Chapter 16 – Drug Lab

When we discuss opportunities for research and development, there remains an unsung hero, the Armed Forces Medical Examiner System, commonly known as AFMES. AFMES provides narcotics training aids to all our military working dog kennels. They are also an investigative, research and development agency. In their collaborations with the services, they conduct canine intelligence—identifying narcotics trends being introduced into society that are potential threats to the military.

AFMES acquires narcotics from the DEA. AFMES configures these drugs into individual narcotics training aids and ships them to military installations worldwide. AFMES, Division of Forensic Toxicology at Dover AFB, executes this joint operation, which is overseen by an Army and Navy medical service forensic toxicologist.

Who has the direct responsibility to fund AFMES? No surprise. It's the DoD Executive Agent. One of the specific elements of the DoD MWD Program Manager's job description is counter-narcotics. While there was a hiatus as we mustered canine forces to support the war in the Middle East, the mission remains relevant. In 2018, U.S. NORTHCOM requested DoD support to provide narcotics detection dogs. The mission—counter-drug deterrence and detection along the southern border supporting Customs and Border Protection. This was well received news, as the counter-narcotics mission in Afghanistan had terminated. Narcotics dogs were returned to performing health and welfare searches on U.S. bases. Drug dogs went from a high operational tempo to being minimally utilized.

The services were eager to comply. Drug dogs from all the services supported Task Force (TF) North missions. Resulting in the seizure of several thousand pounds of illegal drugs. Yet our dog teams were out there unprotected from the threat of fentanyl exposure. This was another battle that was a priority of effort for the Joint Services and Veterinary Community, but one in which the EA had little or no interest.

The threat existed and God knows that fentanyl today is the number one chemical threat to our human and canine first responders. But the problem that existed then—and today—is that there is no training aid that would provide a trace amount of fentanyl. Who could take on that task? Narcotics training aids equals Drug Lab! We reached out and presented the problem to the Lab. Turned out that LTC Dave Santori and Lt (Navy) Ken Lindsay were already looking into starting the process.

This was fantastic news. We brought in Customs and Border Protection (CBP), the team that dealt with this scourge daily. Our initial meeting was at Customs headquarters. CBP outlined the threat and described the issues they faced in the field. They explained their attempts to ensure dogs had the reverse agent Narcan available for the officer and canine to effectively save their lives. CBP became an invaluable partner with the Joint Service MWD Working Dog Committee. They joined our team, expanding our inter-agency counter-drug efforts. For Dave and Ken, this collaboration allowed them to collect technical and scientific data that we hoped would lead to building a fentanyl training aid.

This projected collaboration with AFMES brought further institutional knowledge and canine intelligence. The goal, as with every initiative, is to improve the DoD MWD Program. The fentanyl training aid would not be just a DoD training tool. It would also be advocated for use in the law enforcement community, and for domestic and global first responders. Until 2018, the only interaction the DoD had with the drug lab was emails and administrative actions required to verify annual accountability of the training aids. The Drug Lab could have been located on the moon. We had a huge asset that was lying dormant. Because of this non-relationship, the narcotics training aids inventory had not been changed for years. This was another issue the Joint Services would tackle. The Drug lab became an active stakeholder with the DoD MWD program. Complacency gave way to interoperability.

So how does the DoD pay for this inter-agency support? A view of the process will help you understand how this tap dance occurs or is supposed to work.

A bi-annual Memorandum of Agreement establishes the roles and responsibilities of Armed Forces Examination Service (AFMES) and the DoD Executive Agent (Air Force).

AFMES Roles and Responsibilities

1. Interface with the DEA: Order, receive narcotics training aids.
2. Prepare narcotics training aids packages for annual distribution to MWD kennels worldwide.
3. Conduct analysis of narcotics trends/threats.

DoD Executive Agent (Air Force)

1. Provide two airmen to augment AFMES to build and ship narcotics training aids. Normally assigned from the traditional AFSF squadron on a loaner basis. No special clinical requirements. Airman assigned for two years.
2. Allocate operational costs in the amount of approximately $30,000 annually for AFMES services.

You would think this would be a simple task. In 2014 the AF failed to allocate two airmen to backfill the outgoing airman. Defaulting on the memorandum of agreement. The AF was only able to offer one replacement. AFMES pushed back, demanding the second airman. The Air Force had no intention of providing a second body. The DoD reached out to the Army, Navy, and Marine Corps…and solicited their services to provide a body. This would be the gap filler providing the EA time to figure out how to meet its responsibility… before the Lab shut down production. This was an absolute failure of planning and executive agency management.

If AFMES shut down, every narcotic dog would be in jeopardy of being decertified. All drug dogs would be suspended from duty. The Air Force assumed no responsibility for solving their own internal problem. As is the case today, its program management mantra is to pass the buck and kick the can to someone else. The solution was provided by the Army.

The Army MWD program is unique in that it is the only program that has a Military Working dog unit in the reserve components. The Connecticut Army National Guard's 928th MWD detachment saved the day. They are part of the Army canine force structure, but they are Title 32 soldiers, under the Command and Control of the Governor of the State of Connecticut. They have a dual mission, supporting the State, and when mobilized, placed in federal service. They are manned by full time Active Guard Reserve (AGR) soldiers. These soldiers are full-time troops permanently assigned at the kennels in Connecticut. Under Title 32 they do not change duty stations.

 I was still the Army Program Manager and reached out to the 928th. I wanted to get their input if they would be able to take on this mission. Rotating a handler to Dover on a six-month rotation. The 928th saw this as an opportunity and were immediately on board. Next, reach out to the Connecticut National Guard Plans Training and Operations Officer, Col, now BG Hedenberg. Col "Head" was a multi-faceted strategic mission-oriented career Guard Officer. We had worked together to get the 928th fully integrated into the "total army." Integrating the 928th to support U.S. Secret Service protection missions, be inclusive in Army-wide MWD meetings, and provide opportunities for deployments.

If the Guard accepted this temporary mission, the DoD would provide the funds for the temporary duty, housing, and food. I was keeping my fingers crossed that Col. Hedenberg and the Adjutant General would favorably consider this request. In no uncertain terms, I strongly emphasized that the Guard would be the savior for the entire narcotics dog program. This was not an order. I had no authority as an Army staff officer to direct or order anyone. If their priority of effort was to other State missions, then case closed. This was strictly nothing more than asking a friend for a favor, relying on friendship as much as the need to accomplish this mission.

Within forty-eight hours, Col. Hedenberg came back with the approval…the CTARNG had pulled the DoD's butt out of the fire, preventing the Drug Lab from terminating operations. Within thirty days, a handler from the 928th deployed to Dover, in-processed and commenced the AFMES integration. The 928th soldier, a drug dog handler, it worked like clockwork. He was a stellar performer and temporarily alleviated the failure of the DoD to meet its MOA manning responsibility.

For almost two years, the Guard saved the DoD's bacon. The critical mission finally ended when the Air Force "found" a second airman to fill the Drug Lab position, I believe, September 2016. The Guard, being ever vigilant and mission ready, stepped up to meet a unique mission, turning a DoD failure into a canine handler enhancement opportunity.

While this "band aid" fix was thought to be a onetime event, it was not. When I assumed the DoD position in July 2018, the AFMES – DoD MOA was again coming up for renewal. I collaborated with the AFMES leadership to publish our draft Memorandum of Agreement so it could be staffed. What a surprise… déjà vu… it was not a return to Camelot. History was repeating itself.

The meat of the memorandum was a standard re-statement of the previous MOA with minor technical modifications by AFMES. The two airmen currently assigned were within thirty days of returning to their respective units. As always, our augmentees were exceptional, regardless of what service they were from. The question was where would the backfill airmen come from? A coin toss? No one had a clue. We were moving down that road again—and by October 2018, replacements would have to be on station.

Resourcing these positions was the responsibility of the Air Force Security Forces Senior Enlisted Advisor, a Chief Master Sergeant. I engaged with her numerous times about who would replace these airmen and what would be their report date. She had no idea of the requirements or where they would come from. After receiving a blank stare as my response. I brought the issue to the Deputy Director. Her recommendation was that we provide a line of funding for AFMES so they could hire two civilian contractors to pay the bill. AFMES was initially receptive to that idea, however when the discussion of a timeline for transfer of funds, and who was going to assume contract oversight becoming stalled, that idea quickly disappeared.

The next discussion was the realignment of the Air Force Security Squadron (AFSF) at Dover. This realignment would create excess positions. The first thought for a temporary fix was to take two of those excess positions and reassign them to AFMES. There would be no order or permanent change of station—just change the duty station from Security Forces squadron across the base to AFMES. That never happened. This went on for more than a year with no resolution in sight. Two months before I retired on January 26, 2020, the AFMES bill to provide two airmen remained unresolved. No surprise, typical Air Force Security Forces kicking the can down the road.

Regardless of the EA, the DoD Joint Services moved forward. We opened lines of communication. The drug lab was totally integrated…and besides fentanyl, we discussed creating an LSD training aid. The Lab identified gaps in formal narcotics custodian training for troops assigned as controlled substance custodians. It would include data administration, training aids assembly and basic scientific techniques for troops being assigned to the lab.

What should have been executed was the recommendation by the Joint Services MWD Committee to modify the MOA. Modifying the MOA would task each service to support AFMES with two troops on a two-year rotational cycle. Assignment would be a privilege, and considered a broadening assignment, for selected narcotics dog handlers. The MOA should have been drafted, staffed and attained signatory concurrence from the service Program Managers, then signed by the EA. Unfortunately, the stone age mentality of business as usual let the MOA continue in the quagmire of trying to find two airmen to fill these AFMES positions. Another example where the Air Force intervened, usurping the responsibility and authority of the Joint Services MWD Committee recommendations.

Research and Development Success Stories

Department of Defense Animal Hospital. Probably the biggest breakthrough was the shining study that emerged from the battlefields of Iraq and Afghanistan on canine post-traumatic stress disorder. It became a defining moment in veterinary science…officially verifying the diagnosis of Canine Post Traumatic Stress Disorder (CPTSD). Dr. Walter Burghardt spent three years evaluating traumatized MWDs returning from the CENTCOM theater. He proved conclusively that dogs exposed to traumatic events develop PTSD. This breakthrough opened the door for diagnosis and treatment, rather than the standard practice of euthanizing the dogs. These studies received national acclaim and set the bar for the global Veterinary Community.

Counter Terrorism Technical Support Office (CTTSO): A DoD agency whose primary mission is to conduct the U.S. national interagency research and development program for combating terrorism through rapid research, development, and prototyping. They are independent and not aligned to the bureaucratic research and development denial system found in the military services.

In 2019, I was invited to attend the Countering Terrorism Technical Support Office (CTTSO) project review. I believed that this would be a good networking opportunity to project our Joint Service priorities with stakeholders. What I discovered was that CTTSO was already in a phase II effort with Israel to develop a canine emergency respirator—one of the critical pieces of equipment identified by the Joint Services MWD Committee to provide protection from chemical-biological exposure.

Someone was looking out for the program! It struck me as odd, as none of the DoD MWD stakeholders had been brought into this planning effort. The respirator program continues, with the hope that the final product will be accepted by DoD and approved for procurement. The Canine Handler Advanced Training Simulator (CHATS): An Israeli driven product, this requirement provides simulation-based immersive training technology, such as virtual reality (VR), augmented reality (AR), or mixed reality (MxR) scenarios. Exposing inexperienced handlers to a broad range of tactical decision-making scenarios and dog behaviors that are not dependent on interaction with a real-world MWD. The objective of CHATS is to teach the inexperienced handler how to think critically, improve his/her tactical decision making, and better prepare him/her to train directly with an MWD. The goal is to adapt the simulator to provide better MWD team performance, but also utilized to introduce tactics, techniques, and procedures (TTPs) as a pre-deployment tool.

Training Aid Delivery Device (TADD)

The biggest threat to our military working dogs during the Iraq and Afghanistan wars increasingly became the use of Home-Made Explosives (HME). We had a tremendous shortfall in that there were no HME training aids…and we desperately needed to introduce and imprint our dogs on these odors (TATP and HMTD). This was consistently identified as a priority in our Joint Service MWD Committee meetings.

The Army took the lead, partnering with Army Edgewood Chemical Biological Center (ECBC). Together they would provide the research to develop HME training aids. We also enlisted the help of the Army Rapid Fielding Initiative (REF) whose mission was to provide funding to mitigate needs assessments of identified threats in the combat zone. REF provided the initial startup funding of $500K. The task at hand, finding a funding source to carry on the project to completion. Second, once validated, the long and arduous approval process to add the training aid to the army canine explosive scent kit (CESK).

The next few paragraphs are technical in nature. I tried to cull it down into layman's terms, but the description of the efforts and understanding of the process is important. Development of training protocols was in a secure sterile environment at Edgewood, Md. The test protocols are as follows.

The assembly is only 2.5 inches tall. Scent is released by actuating the flip-top to allow for vapor diffusion. Closing the flip-top markedly reduces scent release. The closed TADD is kept inside of closed zip top mylar bags to further reduce scent release over longer periods of time.

Training Aid Delivery Device (TADD)

Each canine team was subjected to a total of eight sessions on each test date. Each session consisted of two trials. One trial consisted of a canine team searching for a scent carousel for target HME odor. Stainless steel shaker cans contained either target HME odor, non-target items such as distracters or controls, or remained empty to serve as blanks. For each trial, it was possible for the target HME odor to be present or for the trial to be blank, but if the target HME odor was present, only one target HME odor was placed in a single trial. This is the only information the handlers received regarding the location of the target HME odors.

All trials were conducted as a single blind study. The handlers had no information on the location of the targets prior to session searches. Teams were given the option of running on or off leash. The run order was dictated randomly, and multiple passes of the scent carousel were permitted. When a team completed their search, so as not to bias the data, they were sequestered, preventing any interaction with teams that had not yet completed the session.

Scent Carousel

Where is the indelible importance of this R&D? Previously, HME exposure/imprinting/training required a Chemist to mix the highly volatile explosive materials. The dogs would be introduced to the odor and then the material would be destroyed upon completion of the event. TADDS, a self-contained training aid provided trace amounts or odor. Further testing determined that the shelf life of the TADD was approximately thirty days. TADD was not volatile and need not be stored separately. It posed no danger to the handler or dog. The TADD R&D was initiated with the goal of providing a universal training aid for our canine partners in Law Enforcement, and our international coalition partners.

Army MWD teams within the National Capitol Region supported a three-month cycle that provided a qualitative test population and maintained continuity and clarity in test protocols and data collection. The Army led the way on behalf of a joint services requirement. It concluded that TADD was an effective and efficient way of presenting a highly volatile odor. Eliminating injury from pre-mature detonation. The conclusive evidence validated the delivery system for use as a training aid throughout the spectrum of the canine community.

TADD training aids were now able to be transported in a lightweight padded carrying case, commonly known as a "pelican case." It can normally carry between one quarter to four pounds of training aids. This joint effort between ECBC, Army Handlers and OPMG revolutionized deployable training aids, reducing special handling, vehicles, achieving HME training opportunities.

Adopting the protocols used by the ATF, the scientific resolutions allowed trace amounts of explosive odor to be placed in the sealed containers, placed in a pelican case and hand carried. The trace amounts of odor posed no explosive threat, eliminating the requirement to execute time consuming paperwork and carry bulk explosives. Training time improved, without posing a danger to the installation. This also allowed MWD teams to travel to other sites off installation to conduct joint training with other first responders, breaking out of the old paradigms of redundancy.

Chapter 17 – Working Dog Management System

The Working Dog Management System (WDMS) is the computer program that tracks MWDs' data from birth (whelped) to retirement/final disposition. It is used by the services and the 341st Training Readiness Squadron to provide full lifecycle management of military working dogs. It tracks the identity, medical status, training, certification, operational assignment, and disposition of DoD working dogs.

To reach the worldwide customer base, the system is web-based, operating on a framework permitting worldwide access via the Internet and local area networks (LANs). System security through a Secured Socket Layer (SSL) and encryption technologies. Data entry, reporting, and ad-hoc querying can be performed. The application is composed of autonomous modules, operating off a common database. Each module facilitates a business function or using information significant to the lifecycle management of a working dog. This system operates in the Azure Government Cloud under the Federal Risk and Authorization Management Program (FEDRAMP) and is accessible by authorized users with valid access rights, via the non-classified NIPR Net. The system maintains over 20,000 dog records and many other supporting records.

WDMs was designed to allow customers the ability to request portal improvements, ease of access, ability to apply formulas to run multiple reports. As a cost sharing system, each service was assessed a fair share. That provided minimal service portal support. If a service desired specific modifications to their portal, it would be assessed additional program configuration costs. This was another "pass the buck." WDMS was not a joint service system, it was an owned and operated Air Force system, located at Lackland AFB. WDMS system administration was placed on the shoulders of a single WDMS Program Manager, and assistant, a civilian contractor. The Joint Services had minimal input as members of the Systems Configuration Review Board.

In recent years, the Veterinary Community presented a recommendation to merge WDMS with MWD's remote on-line veterinary record (ROVR). This would allow a more efficient and effective systematic interface between the operational canine community and veterinary clinicians assessing canine medical readiness. During my tenure, system administrators designed the interface. However, the process ended up in limbo for close to two years as the memorandum of agreement between the DoD EA and Defense Health Agency continued to bounce back and forth with numerous modifications.

The Air Force WDMS program managers responsibilities were enormous. She managed five times the workload than in her job description. But as she was a Security forces Center employee, the DoD MWD Program manager had no influence on performing a position description review requesting a position upgrade, influencing manning or directing system administration.

This angel received all the services error reports, and served as the help desk, (WDMS did not have a 24/7 help desk). She was overwhelmed, writing programs, running beta tests, and recommending systems upgrades. How did this all go wrong? In 2014, the Air Force Executive Agent determined that they would assume system responsibility for WDMS. With the swipe of a pen, the services lost WDMS effectiveness and efficiency. The Air Force, unprepared and incapable of system management, saw an immediate degradation of system capability.

Prior to 2014, each service contracted their own WDMS administration independently through a government contractor. Each service also allocated funding for their systems management. Contractors provided immediate response and recommended systems improvements. The problem was that the EA did not have centralized control nor data interface. The Air Force independently hodge podged a system without a beta test nor input from the services. Nor did they collaborate with the service contractors to establish an orderly systems transition. The result…what you would expect. We went from feast to famine. The services went from contract efficiencies to the dark ages. While centrally managed, it was a total fiasco in customer support.

To pay for Air Force server maintenance, the services were charged an annual fair share cost of $125,000. Like a home cable system, if you wanted upgrades to services you have to pay an additional cost. The Marines paid an additional cost and hired a Marine system administrator, to their benefit. The Marine portal provided the services to meet their system goals. The other services' fair share funding went primarily for system maintenance. The Air Force, as the system administrator, developed the statement of work, without input from the services. It was a mess with customers bearing the brunt of Air Force oversight failures.

The DoD MWD Working Dog Program Manager is appointed in writing as the system program manager. Unfortunately, while I maintained this appointment, a review of my job description did not have WDMS as a task. I attempted multiple times to address this with my O5, but he blatantly refused to modify my job description. Rather I was admonished for failure to comply with Core values of the Executive Agent. What rock were they hidden under?

Conductivity at remote locations plagued deployed forces in Iraq and Afghanistan. More disconcerting was that during deployments, unit kennel masters suspended handlers WDMS accounts, prohibiting troops from uploading critical performance data. Missions, types of explosives/IEDs, training time, capturing environmental work cycles, detection rates went unreported.

That was prior to 2018, when the Air Force enterprise…along with the rest of the DoD… started to enter the 21st Century, taking on the huge transition from fixed servers to worldwide conductivity through the cloud. I was unaware that upon assuming the position as the DoD MWD PM, that twenty-five percent plus of my time would be executing this transition. The onus of this transition thrust me into the position as a true systems administrator. This was a position where I had little or no understanding…or situational awareness. Fortunately, the team led by the Deputy Director provided me with the guidance and benchmarks to drive the transition.

WDMS operating structure was stripped to the bare essentials. A contract team under Air Force A1 served as project lead, with the "Knowledge Transfer Agreement" establishing the implementation plan, transition timelines, systems security, assess current or alternate servers, customer beta testing and systems acceptance. The challenge was daunting. However, the team adapted and because of the leadership and knowledge provided by the developers, our deputy, and our great team at Lackland, we moved towards global cloud access.

Within the transition team, I was appointed as the Mission Application Program Manager. Our deputy served as the system security ATO. My duties included daily interface with the contract developer (CCE), Security Forces Center System administrator, and the Cyber Security team. As the transition progressed, we began systems vulnerability analysis, determining risk values, generating remediation protocols and fix dates. We mapped the topology of the system, the maze of how data would flow and identified access and access controls, along with ensuring adequate security storage in Azure.

Status meetings, intermediate system function checks to access control, data integrity and system availability were assessed weekly. Under the transition, we changed paradigms. No longer an Air Force executed system, we included the services in transition briefings. Their inclusion added critical data that allowed system improvements, immediately reducing error reports. The system was being built from the bottom up. The old days of the Air Force independently excluding the services were over.

The SAR (Security Assessment Report) provided the Authority to Operate (ATO). No system transition would be implemented without the proper security protocols in place. There were six findings:

1. Lack of audit logs.
2. Lack of limited privileges to change codes.
3. Encryption protections not properly implemented.
4. System backups are not conducted.
5. Failure to adequately test.
6. Default system account.

These findings were assessed as "Low" and did not impact the ability to move forward with system integration and the system beta testing. As we moved closer to the hand off, a unique transformation took place. Historically, at the Joint Service MWD semi-annual meetings as far back as 2012, WDMS would be briefed. The reaction would be unanimous…and the briefer would be pummeled with negative comments. "WDMS was a waste of time, effort, and money!" Stakeholders, at the onset, actively engaged in designing the new and improved WDMS. Their input in the beta test changed system negativity to system effectiveness.

A new dawn, a user friendly, globally accessed system. The mission was done. Now with that accomplished, I again raised the need to have my position description modified, easy fix? You would think so. I had just completed a system transition. A modification to my job description should have been a "paper exercise." My request was received with a NO. It was not going to happen until I figured out how to procure and produce dogs for the services. Back to the elusive core requirements.

That comment was ludicrous. I didn't train or procure dogs. That was the 341st Training Readiness Squadron. There was absolutely no reason to deny this modification, but I knew what this was all about. This was another Air Force ploy to put me in my place, to ignore the obvious. However, I had more pressing concerns about some pending surgery to worry about. So, I recorded everything. WDMS was a major undertaking! Through a united effort, we made dramatic changes to system efficiencies, utilization, and acceptance by the community of stakeholders.

Chapter 18 – Inter-Agency Stories

There are so many stories of working with canine partners that it would be impossible to capture all the great work that's being done to build relationships. We had a single focus, to be prepared for the next possible threat to the United States, home or abroad. These are a few of of their stories. The stories of caring, dedicated canine professionals. Just a spattering of the inter-agency canine family that have been invaluable to keeping our communities and nation safe. This is a tribute to what you have done and what you continue to do.

Camp David

With the update Air Force Instruction 31-126, we took the bull by the horns. After years of exclusion, the re-write brought our DoD Civilian Law Enforcement Agencies under the DoD umbrella: Pentagon Force Protection Agency, National Security Agency, National Geospatial Agency, and Defense Intelligence Agency. This unification did not hinder these agencies from executing their separate and distinct missions, what it did was codify their representation under the Department of Defense. It allowed these agencies to become part of the DoD canine family and part of the Joint Service MWD Committee. They benefited from government veterinary resources, research and development. Their assets were fully captured to support external and internal DoD missions, affording them mission opportunities outside the boundaries of their respective campuses.

The consolidation of canine inventory was quickly realized, sooner than anyone thought possible. In the summer of 2019, I received a call from the Marine Corps Security Detachment at Camp David Md. Yes, the Presidential retreat. Camp David is a Marine Corps Installation, and when the President is not in-residence, security of Camp David is a dedicated Marine mission. When POTUS or other designated international leaders are in residence, the Secret Service is the security lead. Marines' security remains in place, taking direction from the Secret Service. After an informative discussion, I learned that when not occupied by POTUS, the Secret Service *did not* provide canine support for the installation. And no Marine MWD teams were permanently assigned. Camp David at the time was relying on the only explosive canine assets available, Frederick County Sheriff's Office. They provided random canine patrols on a handshake basis. The Marine Commander understood this was not the best course of action but was reliant on what was available to accomplish the mission.

With that information, I reached out to the Marine Corps MWD Program Manager, and he confirmed that the Marine Corps did not provide MWDs on a permanent basis to Camp David. Any MWD support for Camp David would have to come from the Marine Detachment at Quantico over 100 miles away. That led me to see if I could ramp up another service to provide interim support. The Military Service Program managers were unable to assist, operational tempo was high and all MWDs were committed.

As we recently brought on the four Civilian DoD Police agencies, I reached out to our civilian law enforcement with the largest canine inventory, the Pentagon Force Protection Agency (PFPA). I presented the mission, and PFPA with thirty canines in their inventory jumped at the opportunity to support. In a matter of days, PFPA and the Marine Detachment had worked out a random-access schedule, with PFPA teams on site for two or three days, on a reimbursable basis. The inclusion of the DoD law enforcement agencies providing additional certified dogs teams was paramount in allowing PFPA to coordinate this very important mission. PFPA stepped outside their box to provide canine explosive detection protection for the Presidential retreat. No fanfare, no recognition.

With PFPA as a DoD partner, it also opened the door for the services to conduct joint training, which included being introduced to the Pentagon Subway station, working in a confined space. As well as acclimation training on their Personal Borne Explosive Device canines, dogs that track a mobile manmade explosive. We were off and running, totally integrating a one team concept that had been alluding to the DoD for years. The NSA, DIA, PFPA, and NGA's teams became tremendous force multipliers in the National Capitol Region, unifying our effort to deter, detect and protect.

NYPD Transit Bureau

My introduction to NYPD transit was through my other dear friend, who I've worked with in NYPD Counterterrorism since 9/11, LT. (ret) Brian Corrigan. I'll talk about Brian later, but through these friendships cultivated our unified effort crossing the lines of jurisdictions with the single purpose of protecting our great nation.

During the TEDD transition NYPD Transit bureau came to North Carolina to evaluate potential canine candidates for their program… protecting the New York City subway system. Their dogs operate in confined spaces, in high density populations, in an environment laden with numerous odors, and carbon monoxide. The NYPD Transit dogs temperament, detection capabilities, reaction to multiple distractors make the selection of the right dogs crucial to perform in this challenging environment.

NYPD Transit Bureau selection criteria is one of the most challenging in the canine community. They screened multiple dogs and selected one…Cesar who had deployed to Afghanistan twice. Cesar was brought back to New York where he was paired with Police Officer Pete Rodriguez, a former Army Ranger. A perfect match. Pete and Cesar completed training in 2014 and for the next six years protected millions of citizens who traveled in the steel tubes of the subways under the city of New York daily. Sadly, Cesar developed cancer and on 15 Feb 2019, Cesar and Pete took the final walk of honor with NYPD canine officers lining the street in tribute.

Police Officer Peter Rodriguez and Cesar (permission for use from P. Rodriguez)

Cesar served both his country and the City of New York honorably and with distinction. Pete and I quickly became friends, and when I would travel up to the city to visit family, I would always stop by the Transit Bureau and check up on my battle buddies. True heroes.

Lt John Papas, SGT Randy Brenner, Detective Scott Spectre, Detective Wayne Rothchild of the NYPD Transit Bureau Canine Unit are recognized global canine subject matter experts. Their devotion, mutual respect, vision, and desire to share their subterranean canine protocols were instrumental opening the door for an NYPD – Military collaboration. It allowed the military to experience in real time canine operations, from NYPD officers, all veterans of the attacks on the World Trade Center.

Pizza

In the summer of 2015, I was in New York visiting my kids. As it happened, I was near the Transit Bureau's "Base." I decided to stop by and see my NYPD brethren, unfortunately LT Pappas was on leave, but SGT Brenner was there, and as always, I was welcomed with open arms. After exchanging greetings and saying hello to the officers, like a canine, I immediately picked up the scent of a smell that all New Yorkers are weaned on…New York Pizza! Randy had ordered a "Pie" and offered to share.

We adjourned to his office and as we sat and chatted, the subject turned to how Counter Terrorism Technical Support Office (CTTSO) was conducting HME training, the training exercise…Flat Rock was going to be in Boston in June. Unfortunately travel outside the limits of New York State was prohibited by the NYPD. While this was a departmental decision, New York remained number one on the hit parade for acts of terrorism. Regardless of Flat Rock being DoD funded the NYPD travel restrictions were set in stone. Could we find a work around? We did.

How and where could we expedite another venue here in New York. So, as we noshed on our "slices," I called the lead for Flat Rock, Ms. Jenna Gadbery. If anyone could figure out how to make this happen, it was Jenna. Innovative and knowing how to maneuver the system, we hoped that she could find funding for the exercise one more time in New York. The call to Jenna started the wheels turning.

We needed "A Favor." Jenna was a friend of the NYPD as was her partner Dr. Michele Maughan. They had birthed TADD and were attempting to provide the opportunity to attend Flat Rock to as many law enforcement agencies as possible. Jenna pondered the concept for a while, and to our surprise informed us that she believed that they probably had sufficient funding to support a Flat Rock exercise at West Point, fifty miles north of the City. Next step was for me…as the Army Program Manager…to submit a request and coordinate training sites with West Point. My next phone call was to the Provost Marshal's Office at West Point…asking for another "favor." The train was picking up steam.

We made contacts with the West Point Director of Force Protection and Range Control. Our first hurdle was that we were looking at a date near the end of July. That same period, the Corps of Cadets floods every training area for Cadet Basic and Cadet Field Training. But we were lucky. There was a window of opportunity for one week, when the stars aligned. Jenna penciled in the Flat Rock Exercise.

It was confirmed. The key to success "Pizza!" Always keep an open mind, never say no and if an opportunity presents itself, take advantage. My focus should have been on what was in the best interest of the Army. But strategically supporting the NYPD was an opportunity for our continued joint efforts to train with the police agency in the city that is still number one on the terrorist hit parade. If the shit hit the fan in New York City, our MWDs would be supporting the NYPD.

Flat Rock - West Point

Flat Rock was unique in that it was a three-phase course. First, handlers were introduced to a training scenario. The scenario was a "blind" (Handlers would not know where the explosive training aids were located). The scenario would challenge their skills in reading the changes in their dogs' behavior in a real-world environment. CTTSO would gage their skills to work a search pattern, while at the same time, introducing Home Made Explosives in pure and pseudo form. Flat rock was also a platform to gather data for scientific analysis of the effectiveness of real and pseudo-odor.

We set up in two areas, one in Giles Field House where the teams received their indoctrination and were run "cold" through the training lanes, first with an evaluation team making notes on critical search techniques, hits and misses. Once that was completed, each handler's data was analyzed. The next phase was one on one training. The team was introduced to an odor in a relatively small search area where the instructor/evaluator could work one on one with them. These engagements were not graded. This opportunity was unique in scenario-based training, as normal exercises terminate in a graded exercise. Flat Rock was non graded, focusing on individual team improvement and the results were shared with the handler. Nothing was provided to their respective units.

When we talk about collaborations, an immense amount of credit must be shared with the Connecticut Army National Guard. Their director of Operation at the time, Col. Ralph Hedenberg (now General) was instrumental to the success of Flat Rock. Instructors and evaluators from the Flat Rock Team relied on canine units to provide explosive training aids. The closest Army unit was the 928th MWD Detachment (Small). With Col. Hedenberg's blessing the 928th provided explosive training aids and received the benefit of participating in the Flat Rock venue. Army led the way!

We had an excellent turn out. Approximately sixty teams from the surrounding counties law enforcement agencies participated. However, most attendees were NYPD transit bureau. We had sixty percent law enforcement and forty percent Military. The value of this training was not measured in attendees. Its value was measured in the multi-agency collaborations.

We added to the training scenario by offering NYPD teams an opportunity to run through the Army certification protocols. MSG Sean Shiplett, one of the MWD Program Managers, was on hand to add his expertise and guidance, as well as to provide exercise feedback. Sean had years of experience, with several deployments. He jumped in to share that experience with our law enforcement partners and run them through Army certifications. Army certifications were more stringent in time and area. Some took close to an hour to conduct in an open area or vehicle search (up to 100 vehicles). The certification scenarios paralleled the conditions military teams would encounter in combat operations.

At the end of the day, Flat Rock and the secondary team effort to provide law enforcement an opportunity to train…and attain Army certification… was a resounding success.

National Odor Recognition Test (NORT)

Congress appointed the Bureau of Alcohol Tobacco and Firearms to lead the effort to introduce asymmetric explosives to the law enforcement and military community. Annually, the ATF deployed across the nation and presented a three-day certification course. This was a unique opportunity to continue what we accomplished at Flat Rock.

The ATF had been imprinting our deploying dog teams at Yuma Proving Grounds, AZ. NORT achieved the same mission, with the twist that NORT awarded a certification. Same scenario, introduce and certify dogs on TATP and HMTD. But like Flat Rock ATF trainers worked with handlers to help hone their skills. A win-win.

The ATF provided us with their annual schedule and gave us an open invitation to participate in NORT certifications. Initially, military teams participated as the opportunity existed. But there was a $30K agency cost to schedule dedicated NORT certification courses. Both the Army and Marine Corps saw the need to capitalize on NORT. It accomplished several instrumental goals. It gave our handlers the opportunity to attain a national certification. And it provided another opportunity to join our interagency partners. That always paid off as the exchange of experiences and training techniques was just as important as the actual formal certification. We ran NORT for three years, continuing to capitalize on regional areas throughout the U.S. mainland. We specifically focused on the Northeast, running NORT at West Point for two more years, providing mutual training support for NYPD Counterterrorism, Transit Bureau, and Emergency Service Units plus multiple agencies from the tri-state (NY-NJ-CT) and our sister military services. Costly, but the mission was accomplished.

National Odor Recognition Test (NORT) -Aloha

For NORT expansions, 2017 was a banner year. NORT had never been conducted in the Pacific. Transportation of the volatile training aids and travel costs played into the decision of not conducting NORT in Hawaii. On the other hand, the need to get out there was mission essential. Finally, through some creative financing, the ATF worked through TSA to attain transportation procedures to move the explosive training aids… we were off and running. MSG Lavalle…the U.S. Army Pacific MWD Program Manager led the way. She collaborated with the Air Force, Army, Marine Corps, Navy and, Honolulu Police Department. We were fortunate that our Pacific partners from Australia and New Zealand accepted our invitation and confirmed their attendance. This was money well spent. This was another initiative where our great NCO team worked the wickets to expand mission essential training and global outreach.

FBI – Explosive Division – Saint Kelly

I've touched on the Joint Memorandum of Agreement that we achieved in 2019. And I alluded to the dilemma and actions that were taken to approve it. A document that glued the FBI as a permanent fixture to the DoD MWD program. An enduring partnership solidifying Home Made Explosives exposure to our MWD's.

By 2019, the $30K cost for the ATF NORT program stretched Marine and Army budgets… NORT was terminated. Our deployments dramatically curtailed the need to send dog teams to the Marine Pre-deployment training center. However, the reduction of contingency deployments did not eliminate the need to train our military working dogs to detect homemade explosives. HME had been used in attacks against civilians in Germany, France, and Belgium. The four-letter curse for MWDs… TATP and HMTD.

The solution came from my ever-present guardian angel, and friend, Dave Kontny, Chief of Staff, Joint Program Office of the FBI. Dave advised me that the FBI Explosive Division was conducting training exercises to introduce homemade explosives to civilian police agencies throughout the nation. He recommended that we get in contact with their lead agent, Kelly.

Along with my partners, Jimmy and Jason, we made arrangements to meet Kelly at the FBI Explosive Division in Quantico, VA. The FBI program was an "acclimation" training event that would not offer a certification. They would be able to support training on the mainland and overseas in Hawaii, Guam and Puerto Rico. This would cover down on all kennels minus our MWD kennels in Europe. This was not an issue as the Belgian Federal Police supported our European kennels' homemade explosive training.

Our meeting was transformational. Kelly jumped right in. The FBI asked how they could help. We were even more surprised when Kelly informed us that this training would be offered to the military without cost. From our conversation she started to ask specific questions to frame her plan —kennel locations, kennel masters' names, numbers to be trained, etc. In quick succession, she provided us with her concept plan. In one afternoon, we forged an inter-agency partnership. The Army…soon to be carried over to the DoD…achieved our HME readiness because of the spectacular efforts and friendships with the FBI.

In July 2017, we conducted the proof of principle at Ft. Jackson, South Carolina. SSG Nick Briggs, the kennel master, hosted the event. The only cost to the Army were the cinder blocks they'd procured as a safety precaution, placing the TATP and HMTD in the cinder blocks to absorb any premature detonations. FBI Special Agent Bomb Technicians (SABTs) across the nation were our liaisons. They procured, mixed, and set the training aids.

FBI protocols, as opposed to the ATF, had no time restrictions. Once the explosive component and safety briefings were given, dogs were introduced to the odor. Proficiency training was repeated as many times as the team felt was necessary. Training was effortless. FBI Special Agent Bomb Technicians (SABT), kennel master, trainers and MWD teams worked seamlessly. The FBI SABT professionalism and commitment to "taking care of their troops" was an honor and blessing for our MWD teams. The event achieved one hundred percent success. The Army executed a Memorandum of Agreement that solidified their training relationship.

What was needed next was to expand the program to cover the entire DoD. The stories, trials, and tribulations to execute the DoD MOA are well presented in the previous paragraphs. As a close-knit community, how do most initiatives get started? Through friendships and favors. We achieved the mission by working hand in glove! What I learned… to my chagrin and have noted throughout this book…is how Air Force EA tunnel vision and claims that our initiatives were "mission creep" was institutional culture. Pass the buck and punt. Ingenuity, perseverance, and strategic maneuvering achieved the mission, despite the EA. Dave, Kelly and all of the SABTs…to this day…should be sainted. These heroes are now an integral part of the DoD MWD team, providing a training venue that prepares our teams for all contingencies, domestic and overseas.

National Explosive Canine Standards

Presidential Policy Directive 17, *Countering Improvised Explosive Devices*, guides our nation's efforts to counter the use of IEDs. Methodologies to protect the United States, our allies and partners, and our interests. PPD-17 directs U.S. government departments and agencies to establish and implement measures to prevent, protect against, respond to, recover from, and mitigate IED attacks and their consequences at home and abroad. The United States recognizes the complex and transnational nature of the IED threat and focuses its counter-IED policy on two primary goals: discovering IED threats early and synchronizing interagency capabilities to address them.

The FBI's Joint Program Office for Countering IEDs became the Department of Homeland Security's lead for this national policy. Initially, the panel consisted of the five Homeland Security Agencies, AMTRAK, and the Federal Protective Services. The DoD was invited to sit on the national canine detection dog standards panel, but only as an observer. While this was a good collaboration of subject matter experts, local law enforcement was not included on the panel. Dave knew the exact agency to include—the NYPD.

When NYPD established their Critical Response Command (CRC), one of the units stood up was the Counter-Terrorism Canine Unit, commanded by Lt. Brian Corrigan. Brian and I went back to 2002. At the time, I was teaching Maritime Security at the U.S. Merchant Marine Academy Global Maritime and Transportation School. The NYPD was expanding their efforts to ensure that they were prepared to provide security for the multi-modal transportation threats. We made that happen and over the last twenty years, we've worked together. A bond between cops.

With Dave's concurrence, I broached having CRC Canine participate on board the standardization team. It only made sense. Paws dipped on the streets of the largest police department in the nation, whose resilience during and after 9/11 drove counterterrorism actions, should be a member of the team. Brian ran it up his chain to his Chief, who saw not only the relevance but the critical importance of inclusion. The need to get the view from the "street," brought efficacy and relevance on behalf of local law enforcement.

Brian's thirty-five years of knowledge in a densely populated Mega-City, that posed multiple odor distractors and multi-cultural perceptions of dogs, helped present a unique perspective. It was approximately eighteen months from the time Brian came onboard to the time the American Academy of Forensic Sciences and American National Standards Institute (ANSI) published the standards. The culmination, "Standard for Training and Certification of Canine Detection of Explosives, 092," was published in July 2021. It was a magnificent effort that provides one set of standards for all explosive detection canines across, not only the U.S. canine enterprise, but our international law enforcement and military canine partners.

The NYPD – NATO Partners

The NYPD has led the way with integration and implementation of multiple counter-terrorism strategies since 9/11. NYPD also employs international liaison officers across the globe working with counter-terrorism partners. That partnership…also includes NYPD hospitality to international canine partners.

The NATO MWD Working group consists of thirteen international partners. Their mission, since its inception in 2010, has been to collaborate and develop NATO procedures that project unified interoperability. Our partners always voiced their desire to visit New York, to visit the 9/11 memorial and meet the NYPD and FDNY heroes, who have been the tip of the counter-terrorism canine operations' spear.

The NATO panel rotates meetings between partner countries. For 2016, I nominated the panel meeting in New York City. The recommendation was unanimously approved. Our meeting would be hosted by the NYPD Counterterrorism, Transit Bureau, and FDNY Fire Marshals Canine units. NYPD also provided the security for the conference. SGT Randy Brenner NYPD Transit, Chief Shanley and LT Corrigan NYPD Counterterrorism, provided a down and dirty briefing on the employment of their canines, paralleled with an in-depth threat analysis. Fire Marshal Joseph DiGiacomo (USMC retired) presented an overview of the FDNY Arson dog program, a capability many overlook, but another critical asset in the war on terror. This mission briefing on the tactics, techniques and procedures was eye opening. The intricate ballet needed to integrate canine operations in a city of over eight million people grabbed everyone's attention.

For our NATO partners, having the opportunity to visit the 9/11 memorial was overwhelming. Every NATO representative had deployed to Iraq or Afghanistan. They knew they were at ground zero of the global war on terrorism. They all were serving their own nations because of what happened in New York City. They understood the impact of what happened there, and they knew they were standing on hallowed grounds, where so many perished. It was a humbling experience.

In the coming years, because of this collaboration, the NYPD hosted military and law enforcement canine handlers and kennel masters from Germany, the UK and Australia. Not just to do a meet and greet or receive the standard canned briefing! No, they wanted to see and learn firsthand. Integrated with NYPD canine teams, they went out on patrol to observe. They asked questions and shared their canine experiences. They learned! Preparing for the next major catastrophe.

The clarity and interaction of these visits cemented new relationships with our European and Pacific Rim civilian law enforcement canine partners. This led to Deputy Chief Shanley and Brian open invitation to attend European Union (EU) counter-terrorism conferences in Belgium. Doors were opening.

Trump Towers – FDNY– New York's Bravest

Post 2016 Presidential election, the Secret Service designated Trump Towers in New York City "White House North." This designation mandated that the U.S. Secret Service assume protective responsibility. The military was part of the Secret Service protection package. VIPSA provided Explosive Ordinance and the services assumed MWD responsibility for the interior and exterior of the building, primarily at the vehicle check point on 56th Street and Lexington Avenue. It also required coordination with our NYPD counter-terrorism partners as the public sidewalks around the building were NYPD jurisdiction.

For the DoD, the first step was to determine how many dogs and how long they would be deployed to New York. Bottom line, if we didn't figure that out from the DoD side, the U.S. Secret Service would direct the tasking. Discussions with the Secret Service and the Air Force Tasking Cell led to a sound rotational plan that would maintain continuity and provide dogs based on a mission analysis. The Secret Service requested twelve teams. DoD stood their ground, and our mission analysis determined the right mix would be seven teams, split between the services on thirty-day rotations. The Secret Service provided the remaining five teams. That was the easy part.

While a thirty-day rotation provided continuity, dogs still needed to get on odor to maintain proficiency. The USSS did not have training aids to support the military teams. Service canine regulations stated that if dogs do not get "on odor" (training) within thirty days, their certifications could be suspended. The first solution was to send teams to the kennels at McGuire AFB for mid-tour training (fifteen days). These dog teams would travel sixty miles to McGuire for a day, get on odor, and use the obedience course to maintain proficiency. The issue —should there be a threat or incident at the Towers, these teams would not be immediately available. We fixed one dilemma but created another.

Calling on our NYPD partners, we tried to collaborate with NYPD Transit/ Counterterrorism to use their training aids. They were supportive. However, their OPTEMPO, trying to provide canine support 24/7 across the city, made it almost impossible to carve out time or dedicate a trainer to help our MWD teams. However, when one door closes another one opens. Fire Marshal Joseph DiGiacomo had recently been selected to stand up the FDNY Arson Dog program. The FDNY canine unit was based at the FDNY Fire Training Academy on Randall's Island, just blocks from Trump Towers. A sprawling training facility, the Academy, is a city within itself. It includes mock high-rise buildings, vehicles, mockups of a ship and even a subway station. Joe is a gentle, kind man, driven to get his mission done. He is also a Marine, a hard charger, a Marines Marine. All it took was a phone call and Joe stepped up to support the Trump Tower mission.

Arson dogs' primary mission is to detect accelerants. But their training includes detecting multiple explosives. Joe's training inventory included the full complement of military explosives. Joe was the savior, making sure our dogs got on odor, and his efforts removed the process of going to McGuire that was neither cost nor mission effective. Fifteen minutes to Rikers Island, teams could easily be recalled to Trump Towers. Did the Secret Service or DoD need to affect a Memorandum of Agreement, was there any "staffing actions" required? A phone call and handshake, the way we do business in the canine community.

One thing I can tell you about the Marines, boot camp makes you rank conscious. Enlisted personnel salute officers, officers return the salute, acknowledging that gesture of respect. The key to the process is that you salute the rank, not the person. A good officer realizes that you must earn that respect. I met Joe in 2016. I was introduced as Colonel and since that time, he has never called me by my first name. To this day, he has yet to refer to me other than by my former rank, "Colonel." We've been friends now for seven years. I hope that he calls me by my rank only because I "earned his respect."

However, Joe…having deployed multiple times…realized that most of the military teams had not been deployed. He saw it as his responsibility to "train them up." He set up challenging training scenarios, tapping into his own knowledge and combat experiences in urban firefighting. Training at the Fire academy extended our training time, almost six hours of dedicated training opposed to the three, that we got at McGuire. It was effective, eliminating $30 in tolls. This was both scheduled and unscheduled. If dog teams were off and Joe had time on his schedule, they were at Randall's training.

These teams provided rave after action reports. Joe challenged them, running them through hi-rise buildings and an actual subway mock-up. In this way, he exposed them to real life subterranean and building searches. These opportunistic training venues are not offered at home station kennels. Handlers said that this program was so realistic that they recommended it become a pre-deployment training site. Keep that in mind. To prepare for the next contingency, which may be domestic, we need to train as we will be deployed. Where will pre-deployment training be when we are sent to support the next contingency?

New York Leads the Way

Joe and a stepchild from the FBI, a narcotics dog handler, Ryan Christie birthed an idea to conduct a regional canine exercise in the tri-state area (New York, New Jersey, Connecticut.) This resulted in a joint inter-agency exercise that included more than four hundred canine first responders from multiple jurisdictions. This project was hosted at the FDNY Fire Academy incorporating multiple venues:

1. Search areas, i.e., rubble pile, open area searches, vessel searches.
2. Canine first aid training, including chem-bio decontamination provided by a U.S. Special Operations Veterinarian.
3. FBI Home Made Explosive imprinting by Special Agent Bomb Technicians from the NY Field Office.
4. USCG Coast Guard Cutters for vessel acclimation.

Joe and his team established a canine exercise template that should be adopted nationwide. As we continue our reliance on canines to detect explosives, electronics, currency, narcotics, chem-bio, and now medical (COVID), canine interoperability remains more important than ever.

Planning took months of time and effort, never conducted on such a large scale. But the missing link, Senior Leaders. Neither the Mayor's office nor the Office of Emergency Management attended. Not one senior officials from any of the tri-state areas bothered to attend the training venue. Joe and Ryan's event was solely canine team driven. Canine institutional culture, ensuring inclusion in emergency planning. One notable failed opportunity, integrating with the Cities paper-driven Veterinary Emergency Response Teams (VERT). Regardless, from a canine perspective the exercise was a complete success. Training objectives met and hopefully this will light the fire for future inter-agency exercises.

The most important takeaways from After Action reviews indicated the same call for action that has been voiced for years: Inclusion, Inclusion, Inclusion! Joe and Ryan brought together an amazing exercise of canine operators. The short fall; lack of command and control, veterinary support, long range planning and representation in Offices of Emergency Management. Compartmentalization remains the biggest threat to how elected officials and their staff see canines. The underlying political climate of "defund the police" has also served as a detractor for many agencies canine programs. Bottom line…they are excluded from military regional war plans and civilian Emergency plans. Joe, Ryan, and all the first responders pulled off a canine miracle. Lost in the halls of the Pentagon, and City Hall.

Our mission, to drive on, raising awareness and understanding that our dogs' "in service" are not magically trained to do a job. It takes time and effort. The dogs do not wag the logistical tail. That should be the goal of Joe and Ryan's next exercise, unity of effort and economy of force. The trick will be to change the institutional culture of a dog as a pet, to a dog as a detection system. One that is so unique it can't be matched, one that does what a human can't, find very bad things and save lives.

NYPD Heads South

The once multi-million-dollar Joint Improvised Device Defeat Organization (JIEDDO) funding and staff were reduced in 2018, the organization re-designated the Joint Improvised Defeat Agency (JIDA). Their mission remained the same, conduct research and development to counter improvised explosive devices. But JIDA was starting to look outside the box, identifying contributing factors to mitigate IED threats.

In March 2018, JIDA conducted their open house at Ft. Belvoir, VA. I was interested to see if any of their projects were going to utilize MWDs. JIDA was in the process of considering dogs as part of a study to identify explosive devices in a high-density population. JIDA was considering a detection scenario, conducting quantitative and qualitative research evaluating detection capabilities on a moving target. I had a quick conversation with the project manager, agreeing that MWDs would have value in the study. This is a ray of light. We were finally going to have that seat at the JIDA table.

The Army would provide Patrol Explosive Detection Dogs (PEDDs) as our test population. The test scenario would be a Non-Combatant Evacuation Operation or NEO. In this scenario the test population would consist of refugees and dependents being evacuated from a theater of operation. Simulating a human wave of people moving down a main roadway fleeing military action. The goal of the test, to assess detection of mobile suicide bombers laden with explosives intermingled with refugees and dependents. This was unique for military working dogs. The standard search protocol for military dogs was the detection of static explosives (buried, concealed) normally in an open area. Using dogs in this scenario would challenge Army dogs in a non-traditional tracking and trailing protocol. Can Army dogs track a mobile explosive target? Army dogs are not trained in this type of detection protocol. Second was there a canine capability trained to track a mobile target? How would we be able to perform a comparative analysis.

The NYPD transit and counter-terrorism bureau had employed a new capability, an Auburn university patented vapor wake dog. These dogs were specially trained to detect and track mobile targets. I presented a proposal to JIDA recommending that we include NYPD vapor dogs in the test. It would allow a comparative analysis and provide what I believed would be conclusive evidence of this capability and the need to investigate the inclusion of vapor dogs into the military inventory.

I reached out and unfortunately NYPD Transit was overcommitted. I then ran the concept by Lt Corrigan NYPD counter-terrorism canine. Immediately, he concurred that he wanted to participate in the test. Per his guidance, I initiated a request to Chief James Waters, the NYPD Counter-Terrorism Bureau Chief. I was familiar with the Chief. He was an outstanding leader and saw the importance of collaboration. I submitted the request stating the importance of this inter-agency qualitative test. We quickly received approval from Chief Waters.

To be honest, without the NYPD's participation the comparative analysis would not have been achieved and the test would have only provided a single source of data, with inconclusive results.

The exercise kicked off in July 2018 at Fort Eustis, VA. JIDA established their command post on day one, running through technical and equipment exercise rehearsals, ensuring communications, filming and prepping the test site. The test population of refugees and dependents were 100 eager DoD, military, and contractors.

The day prior to the exercise the Army and NYPD dogs' teams met at Ft. Eustis. They got down to business and immediately bonded. eager to get their dogs on odor. Unfortunately, there was a glitch, we believed that JIDA would provide explosive training aids. But wires got crossed and that fell through, Murphy had arrived, and the test was in jeopardy.

Like a gift from heaven, there were other teams training in the area, not part of our group, in civilian clothes geared up, many bearded and salty. They were part of Dev Group, the Navy Special Warfare community. Dog folks seem to understand other dog folks…we quickly started up a conversation lamenting on our problem. The sky opened and Dev Group was more than happy to help and get us the training aids we needed. We avoided cancelling the test. The Army test was saved by the Navy!

For the remainder of the day, we pooled our NYPD, Dev Group and Army teams, running multiple scenarios, learning, and exchanging information on search tactics, techniques, and procedures. The Army and Dev Group were extremely interested in the NYPD Vapor Wake dog teams. Birds of a feather. Averting disaster, inadvertently expanding our horizons of canine partnerships.

Day two, JIDA established the detection protocols. These benchmarks would validate the test detection protocols. In the afternoon we moved to the exercise area, the role players mustered, our dog teams standing at the ready. Upon completion of the exercise briefing, approximately one hundred closely dispersed "refugees" started to move down a pre-designated path leading from the beach to the parking lot. Four refugees had suicide vests with explosive training aids. We ran two teams back-to-back, first the military then the NYPD teams. Handlers methodically introduced their dogs, zig zagging from front to rear of the "refugees." We ran the exercise three times. The results were reflective of their detection training. The military teams missed the targets three out of three times. NYPD teams successfully detected odor on the "suicide bombers" quickly picking up the odor and trailing the target in three out of three attempts. It was a clean test. This is what we needed. The comparative analysis was conclusive. The training protocols for all military service explosive detection dogs confirmed their skill to identify a stationary target. Army dogs were not specially trained in "air scent" protocols. This was neither a testament to the proficiencies or lack thereof of either the NYPD or Army teams.

What the test proved for JIDA was that canines are an important counter-IED detection system. The capability codified was the Vapor Wake dog. From the perspective of JIDA and the DoD, a needs assessment on the possible inclusion of an air scent capability for the military. To this date the EA has not conducted a needs assessment or collaborated with the services or combatant commands to perform canine intelligence on a need for air scent dogs. In retrospect, our Israeli partners have employed air scent dogs for years. Is there a need?

U.S. Coast Guard – Distant Partners

In 2016, the U.S. Coast Guard transitioned from the Department of Defense to the Department of Department of Homeland Security. While retaining their DoD mission in the event of a Presidential nationally declared operation. The transition to Homeland Security changed their operational charter, expanding their role from a dedicated military to a federal law enforcement agency. Their mission statement:

The Coast Guard is the principal Federal agency responsible for maritime safety, security, and environmental stewardship in U.S. ports and waterways. In this capacity, the Coast Guard protects and defends more than 100,000 miles of U.S. coastline and inland waterways and safeguards the world's largest Exclusive Economic Zone encompassing 4.5 million square miles. One of the five Armed Services of the United States, the Coast Guard is the only military branch within the Department of Homeland Security. In addition to its role as an Armed Service, the Coast Guard is a first responder and humanitarian service that provides aid to people in distress or impacted by natural and man-made disasters whether at sea or ashore. The Coast Guard is a member of the Intelligence Community and is a law enforcement and regulatory agency with broad legal authorities associated with maritime transportation, hazardous materials shipping, bridge administration, oil spill response, pilotage, and vessel construction and operation. The over 49,000 members of the Coast Guard operate a multi-mission, interoperable fleet of 243 Cutters, 201 fixed and rotary-wing aircraft, and over 1,600 boats. Operational control of surface and air assets is vested in two Coast Guard Areas (Pacific and Atlantic), nine Coast Guard Districts, and thirty-five Sectors located at strategic ports throughout the country. Six Mission Support Logistics and Service Centers provide services for operational assets and shore facilities. Coast Guard program oversight, policy development, and personnel administration.

With this massive national security mission, with interception and deterrence of narcotics entering the United States, the Coast Guard is sourced with only eighteen explosive detection dogs and zero narcotics dogs. The Coast Guard handlers are trained for explosive detection at Lackland by the TSA training academy. The TSA academy is strictly a handler's course. Coast Guard handlers serve as law enforcement officers. The DoD handlers' course for the Army, Navy, Air Force and Marines is a law enforcement handlers course. Would the mission set be better served if CG handlers attended the DoD canine handlers' course?

In 2018, the DoD reached out to the USCG establishing lines of communication. That started a collaborative effort to build a DoD – USCG canine partnership. As the USCG does not have drug dogs, the DoD concurred that patrol narcotics detection dogs (PNDDs) were not being utilized at a high operational tempo. In the spirit of interagency cooperation, the DoD agreed to work with the Coast Guard to develop a memorandum of agreement (MOA) to provide drug dogs to support Random Access Measure (RAM) at Coast Guard bases. Simply stated, Coast Guard installations would contact DoD installations and independently coordinate, where available, DoD drug dogs to do health and welfare searches, a start point.

Simultaneously the Coast Guard would provide our MWD teams acclimation to "water borne" indoctrination on their vessels, and helicopter hoist operations in their search and rescue aircraft. An amazing opportunity. It is extremely feasible that MWD provide inter-agency searches supporting port security, customs, vessel and cargo inspections. And not to be ruled out MWDs deployed to conduct counter-narcotics, counter-smuggling, and counter-piracy missions afloat.

In fact, Army dogs from the 928th MWD Detachment from Connecticut had supported Customs and Border Protection vessel inspections of arriving vessels in the Harbor of the Port of New York. Also supporting the Coast Guard on domestic passenger/vehicle ferries, performing narcotics interdiction on cross Long Island sound ferries. The primary mission…detection, but also provided information for further MWD medical research: 1) Dogs get seasick, 2) Dogs can suffer from vertigo (balance issues).

In October 2019 this was presented as an agenda item to the JSMWDC. The concurrence from the services was unanimous, but they requested time to review the memorandum and provide comments. Temporarily tabled for the next meeting. My USCG counterpart had conducted the staffing and received formal approval to move forward from their Deputy Commandant. I requested to brief the concept to the Executive Agent. But in typical fashion my immediate supervisory chain refused to allow me access to the EA. Not an Air Force issue. My job description as the Single Point of Contact to the Executive Agent wasn't worth the paper it was written on.

I had the displeasure to inform my CG counterpart that the MOA was being intentionally held and threw the 05, whose personal disdain for the program and me was becoming more obvious, under the bus. The issue was elevated to the senior leadership of the Coast Guard who was going to go toe to toe with the Air Force Executive Agent, whether that happened I am unaware. A chance to drive forward, , train on a mission that in all probability DoD canine assets would perform. Declared null and void due to lack of interest. It turns my stomach that Air Force SF leadership have their heads buried in an air force tarmac. Totally clueless of operational interoperability.

These so-called leaders unaware of the USCG contributions to the Wars in Iraq and Afghanistan. The Coast Guard provided Port Security teams to train the Iraqi fledgling Coast Guard, and HAZMAT teams inspecting all returning cargo from overseas. The only service that had that capability. That the Deputy Customs Program Manager for the entire twenty-six country USCENTCOM Area of Responsibility was a Coast Guard Reserve Chief Petty Officer.

This was not the only initiative that was being investigated by the DoD and USCG. With only eighteen dogs procured through the TSA canine explosive program, they were not trained to execute a military mission. How do we get the Coast Guard and DoD to perform a holistic review of their canine program and pull together a needs assessment. How do we return the CG canine program under the umbrella of the DoD. USCG training with TSA was an agency mismatch. Under the DoD PM they would have input as members of the Joint Service MWD Committee. They would identify annual trained dog requirements. If they wanted to include drug dogs, the 341st would adapt their training program to add to their requirements. Their bill, to provide trainers to train their dogs. However, that would require a change in paradigm and knocking down the Executive Agent brick wall.

Regardless we included the "Coasties" as part of our Joint Service MWD Committee. As observers, their subject matter expertise, operational awareness, and perspectives from canine operations afloat saw a naval dimension to MWD missions.

Readiness and the Future

Chapter 19 – Where We Are Headed

We have amassed lessons learned from multiple canine operations over the years. The program remains stagnated, failing to develop capabilities to meet the challenges on future battlefields (biological, disease, suicide bomber detection). I've already pointed out that survival equipment from exposure to chemical and biological exists for troops, but not for our canine force multipliers. The reality of exposure would be the decimation of our canine detection capability. Where are these concerns in the DoD? Unfunded and forgotten.

This is a story that the DoD does not wish you to hear. It contradicts the happy stories told in Rebecca Frankel's *War Dogs* [Reference 5], Shelia Keenan's *Dogs of War* [Reference 6], Mike Ritland's *Navy Seal Dogs – My Tale of Training Dogs for Combat* [Reference 7] or the great article in the June 2014 National Geographic, *Hero Dog Story* [Reference 8].

These are the good news stories. And there are more. You'll hear about the expertise, devotion and dedication of the Army, Navy, Marine Corps, and the Air Force program managers. You'll see them using knowledge gained from years of service to prepare for the next contingency. You'll learn about the kennel masters, handlers, and trainers—young men and women who place their lives on the line every day. This narrative describes a very small "Canine Band of Brothers" that goes well beyond the confines of our military services. The Coast Guard, European Union, FBI, ATF, Academia, Combatant Commands, Coalition Partners, Defense Health Agency, NYPD, FDNY, TSA, NATO, and 5Eyes (Pacific Partners) are just some of the key stakeholders in a global interoperable program.

However, step behind the glass door of the Pentagon and you'll see that the Air Force as the Executive Agent has no strategic vision. They circle the wagons around a policy that has no relevance for the other military services. Inevitably, the Air-Force controlled EA focuses on the best interests of the Air Force, sacrificing canine readiness for other applications. As a result, what is taught as standard Public Administration theory, a five-year program review, has never been performed. The theoretical foundation is "evaluation equals performance relevance." However, the Air Force has not conducted a canine program review since 1974. They have replaced accepted Public Administration theory with internal management theory.

Our enemies—domestically and abroad—will take advantage of this situation. For example, without canine protective equipment, dogs subjected to chemical or biological attacks, which now include fentanyl as a weapon of mass destruction— or a zoonotic pandemic, would decimate the canine community. If major changes are not made soon, canine force multipliers in the military and law enforcement will become useless. Unfortunately, where we should be—and where we are—remain frighteningly far apart. Ignorance is bliss until ignorance and complacency gets people killed. We cannot forget about Pearl Harbor and 9/11! But we have.

How does the EA deliver the canine message to Congress to obtain support and appropriations for program improvements to meet future challenges? The National Defense Authorization Act of 2017 (NDAA 17) directed the Air Force (Executive Agent) to provide Congress with an annual status report on the health of the DoD program. The report is skewed. Designed by the Executive Agent, it provides only the numbers of dogs procured and trained by the 341st Training Readiness Squadron. It does not address any procurement gaps, the numbers of dogs procured from Europe that wash out, or the current shortfalls of the numbers of dogs the Air Force has been unable to train for more than a decade. It also does not address the millions of dollars wasted in program mismanagement.

The Air Force only provides information for that report that demonstrates "program success." Ask any one of the Service MWD Program Managers about the dwindling numbers of dogs trained for the services. When I left in 2020, the gap was 250. At that time, the EA promised that they were going to pursue a separate contract to close the gap. However, that never came to fruition. Air Force smoke and mirrors.

The report also doesn't discuss the huge gap in competitive fair market value costs. The Air Force is limited to approximately $5000 to purchase a trainable dog. If we dig a little deeper, a cost analysis of the current fair market value shows a huge gap competitive market price. The cost of a dog procured from European breeder's ranges between $12,000 to $30,000. Who are our competitors that come to the table with an open checkbook? Russia, Saudi Arabia, China, our adversaries! Add to this our previous discussion about the lack of research and development! We're in the stone age.

The battle of all battles is associated with the Air Force retaining the 341st Training Readiness Squadron (TRS) as an Air Force run school that provides Joint Services with trained dogs. This is despite their past failure to meet the needs of warfighters throughout the wars in Iraq and Afghanistan. Although the Air Force has the title as a "DoD Joint Service School," the 341st is not Joint. It's run by the Air Force, funded by the Air Force and in the best interests of the Air Force. The military services have recommended that the MWD transition to a joint staffed program, funded under DoD appropriations, be removed from total Air Force control. This is the Air Force…unable to support warfighters, or the services…but holding on for dear life.

Chapter 20 – Canine Leadership Training

One of the major gaps in the Joint Service program across the board is the lack of training for our canine leaders and supported unit leadership. From their introduction to the MWD program as basic handlers, they have one sole responsibility…train to team certification and execute missions. Their role as privates through the ranks of Sergeant and Staff Sergeants revolves around individual team proficiencies. Handlers rise through the ranks, with limited leadership opportunities. The next step in a handler's career is being assigned as the kennel/detachment trainer. They remain with the "dogs" as the position requires expertise in advising and guiding the MWD teams' , working to sustain their detection capabilities, monitoring proficiency, and preparing them for validations and annual certifications.

The jump from trainer is the jump from the fire into the frying pan. They jump directly into the supervisory chain of command. Many times, the jump is difficult as service leadership courses do not address Kennel Master roles. Primary leadership courses are a part of the Non-Commissioned Officers Education System for all services. This is the first step in NCO professional development, including MWD NCO's. Primary leadership courses provide the initial benchmarks of leadership, and most attendees will go onto lead troops as squad or section leaders. How does this apply to the operations of an MWD kennel? It really doesn't.

The services' Non-Commissioned Officers Education Systems do not have an advanced handler's course to prepare appointed Non-Commissioned Officers on the special leadership responsibilities needed to supervise a canine unit. Handlers receive this initial generic NCO leadership training as junior E4s (entry – mid level handlers) with several years of enlisted service. They have no intermediary training until they cross the threshold and assume multi-functional leadership responsibilities. Personnel management, budgeting, mission assignment, mission planning, personal evaluations (evaluation reports), disciplinary actions, troop welfare and accountability, certifications, pre-deployment training, force management, inter-agency collaborations, USSS/SECSTATE missions. And into the fray of installation management, learning the process for major facility improvements and new construction.

Handlers are not groomed for this tumultuous transition. Previously, I discussed the DoD Kennel Master course, an Air Force course developed for the Air Force, conveniently called a Joint Service Course, funded by the Air Force Education Command (AETC). From pre-9/11 to now, the KM course remains relatively untouched. The generic foundations of the course fail in teaching the full spectrum Kennel Master responsibilities. Especially the importance that Kennel Masters will play in contract management.

There are two glaring issues that After Action reports have noted. The first is the lack of contract oversight. Almost every deployed kennel master and Theater Program Manager was tasked to execute working dog contracts. They were responsible for contract performance as an appointed government contract officer representative (COR). Second, the course does not present a block of instruction on Joint and Special Forces Operations. Canine leaders had to learn the concepts of Joint Service integration as force multipliers supporting conventional and Special Operations Forces on the fly.

A glaring gap after years of deployments is the interoperability between EOD and canine teams. After Action reports and lessons learned from contingency and USSS missions show two organizations that should be attached at the hip but remain polarized. In multiple trips to the Iraq and Afghanistan, there was no EOD—MWD team unity. Clearly our dogs are the first responders. The mantra "Canine Detect—EOD Renders Safe" has yet to be a lived experience. We consistently work separately and apart. The first time EOD and canine come together is after a dog positively identifies an explosive device and EOD arrives at the incident site.

This is a Team! We are one entity collectively executing counter-explosive/IED missions. When a dog team has a positive response, are we knowledgeable enough on the basics of explosive ordinance to be able to identify and report the EOD Five W's? Specific EOD checklist information: what has been found, the description of identifying markers, wires, shape, and size? Handlers are trained on explosive components as part of their detection training. Kennel masters have responsibility for accounting for explosive training aids. But we don't maintain a training or operational relationship with our EOD partners.

We need to profess EOD—MWD team consistency and continuity. We should be exchanging After Action reports, working to conduct joint training exercises, and developing a strong EOD-MWD relationship. This is mission essential and could be the difference between life and death. How? Our dog teams are trained on commercially procured explosives. Our adversaries do not believe in using off the shelf dynamite and TNT in a standard ¼ to 1 pound block. There is a difference between commercial and theater explosives. Our adversaries spend countless hours trying to configure explosives with masking materials to throw the dog off scent. In Afghanistan, trying to mask as much as 500 pounds of Home-Made explosives.

It was in Afghanistan that we first learned about volatile peroxide-based explosives (TATP & HMTD). EOD first tested these Home-Made Explosives to assess their components. Fortunately, the canine community…through our partners in the Bureau of Alcohol, Tobacco and Firearms…provided these "theater" explosives to train our MWDs during pre-deployment training. Why is it so important to unify and standardize? Because each of our services has EOD and MWD teams. By working together, we can exploit explosive threats, better understand EOD and MWD TTP's and develop standardized reporting protocols to save time and lives.

Another shortfall is the lack of inter-agency professional education opportunities for our military working dog troops. How will we work with Civil Law Enforcement, Offices of Emergency Management, civilian veterinarians, and public works as we start shifting to a civil defense role. Most of our kennels already have "unofficial" relationships with civilian law enforcement counterparts. Birds of a feather. It's just common sense. In many cases, military installations already have support agreements to respond to active shooters and provide back up support.

Mid-level officers are provided enrichment opportunities to work in private industry and attain Congressional and other governmental fellowships. These opportunities enhance their knowledge and skills, grooming them for senior level leadership. What opportunities are available for mid-to-senior Non-Commissioned Officers? If any, they are extremely limited. Yet we expect our NCOs to understand the global enterprise, placing them in positions where they will be relied upon to be fully interoperable in the industrial-military complex. They need to be educated to be able to perform in a strategic environment.

How do we change that paradigm? We started that process, although unofficially, with our MWD Program Managers in U.S. Army Europe. Retired Master Sergeant Den Abdon and Mr. Duane Stinson (retired Air Force handler) became the Army's voice of the Office of Provost Marshal General for our European Union Civilian Law enforcement community. They forged partnerships that exist today, by collaborating mutual training efforts and being included on the EU counter-terrorism team. But these were "additional duties?". What should be codified in the NCO education system as a perfect match for an enhancement tour. The New York City Police Department has three canine units, each having a commanding officer, but no Kennel Master. The NYPD is the premier law enforcement agency in the nation. They wrote the counter-terrorism playbook. MWD NCO future leaders should be integrated, not only with the NYPD, but all the major metropolitan departments. Not as handlers, but in positions in the command staffs of counterterrorism, intelligence, community affairs, patrol. Learning strategic integration and utilization, learning integration of canines on the same scale as they will experience in a theater of operation.

We already have established canine partnerships with our NATO and European partners, and we should develop exchange programs with our allied partners.

We operate in a global enterprise, yet we continue to keep our canine subject matter experts locked within the confines of their service kennels. The other side to the coin is that as our Kennel Master's progress, once they become Program Managers, they transition into the world of policy, program management, budget, planning and, most importantly, evaluating and recommending the strategic analysis of their MWD services programs. Providing them the opportunity to gain experiences outside the confines of their services will allow these noncommissioned officers opportunities to broaden their horizons and knowledge of global canine interoperability.

How do we get there?

The DoD should grasp this opportunity to develop an outreach for civilian enhancement/intern programs, centrally collaborate with the Service Program Managers and establish a selection criteria and Joint Service selection board. The DoD in collaboration with our NATO, Allied and Civilian Law Enforcement Agencies can identify opportunities for canine exchange programs for Handler, Trainer, Kennel Master, and Program Manager executive levels.

The DoD should perform a program review and totally revise their current kennel-master course to present relevant courses based on current and future MWD roles on a global platform, including contract management, force management, Joint operations, and Secret Service Liaison Positions. Simultaneously, the service NCO education system should be modified to present a leadership track for career-tracked Army/Navy/Marine MWD NCOs.

Chapter 21 – Procurement

Most U.S. private venues and the Military's primary procurement source of "green dogs" are currently from Europe, mostly breeders in the Netherlands and Germany. The cost of procuring dogs from European suppliers and availability of quality dogs due to the explosion of demand from other nations is rapidly dwindling. It is anticipated we will reach the breaking point within the next five years. In 2023, we remain on the precipice of being unable to provide animals to support law enforcement and the military.

The solution is a return to "Buy American." It would require re-establishing and reenlisting domestic breeders, essentially, re-birthing the WWII "Dogs for Defense" program. WWII Dogs for Defense program encouraged private citizens to donate dogs for the war effort. Citizens donated over 20,000 candidates. If we were to return to domestic procurement partnering with the American Kennel Club, the MWD shortfall would be erased and we would never again have to deal with a canine shortage, or supporting a surge.

As I've pointed out, the Air Force has no influence on the MWD program outside the halls of Congress. Buried on the fifth floor of the Pentagon, minus the annual Congressional Whitewash report, they do not have any engagements with the House or Senate Armed Services Committees. Yet in the halls of Congress, there is a Congressional Veterinary Caucus, working to enhance Veterinary medicine. All except the Defense Health Agency are excluded from Caucus membership. And oddly enough, there is no Congressional Caucus for canine. The Department of Homeland Security Congressional Sub-Committee represents their agency, with a direct line of appropriations that provided millions to build the TSA program. Influential in obtaining funding for research and development, with Auburn University. The military remains aloof in their tunnel vision, accepting complacency. The Executive Agent does not foster Congressional engagement, and unfortunately dogs—besides being good photo ops—are a commodity with no political influence.

What the Direction? Where are We Headed?

The aftermath of 9/11 placed a heavy demand on counter-terrorism resources. The market for explosive detection canines (EDCs) surged dramatically, placing a strain on the primary procurement source of trainable "green dogs." Estimates were that the demand increased by over eighty percent globally. Historically, most democratic nations…primarily the United States…relied on European breeders to procure high-drive quality dogs (German Shepherds and Belgian Malinois). Their demonstrated drive, conditioning, longevity, and aggressiveness make them ideal candidates for the police and military. They remain the standard bearers, providing the "template" for canine detection work. But demand has outpaced availability, taxing European breeders to the breaking point. At what point will the U.S. realize that the European cupboard is bare?

Under these circumstances, how long will European breeders remain a viable procurement source? Estimates previously discussed identify the timeline for an exhausted pool of available animals as 2025. Post 9/11, nations that had shown little or no interest in canines went into canine buying frenzy. Competition soared. And European breeders, naturally, capitalized on the skyrocketing demand, selling quality dogs to the highest bidder. High end customers—Saudi Arabia, China, and Russia—realized the importance of explosive dogs and quickly procured green dogs in bulk. The influx of these nations into the market changed the playing field, as they came to the table with open check books. Some purchased dogs for thirty thousand dollars (30K), which quickly tipped the supply and demand scale, not in favor of the U.S.

As far back as October 12, 2005, most participants—all MWD handlers, trainers, and Kennel Masters—concurred that the "DoD needed to widen its vendor base, to add additional procurement venues, particularly returning to U.S. breeders under the guise of "Buy American." They raised further concerns that—as the Air Force is not a Joint Services ITRO school—it likely retains the best dogs for its own service. The less than stellar ones go to other services. This perception is a valid argument for re-evaluating the EA to return to a Joint Services program under the Office of Secretary of Defense (OSD). This process should be a unified joint service protocol to ensure total transparency. An independent analysis in Michael Hammerstrom's, 2005 Naval Post Graduate School Thesis, titled: <u>GROUND DOG DAY</u>: LESSONS DON'T HAVE TO BE RELEARNED IN THE USE OF DOGS IN COMBAT, codified the exact same concerns as this author. Remember that Hammerstrom was not involved with Military Working Dogs:

> *"...the DoD needs to widen its vendor base and look at other vendors, particularly in the U.S. The personnel they have selecting dogs for training must realize that we are looking for War Dogs and not police K-9s. The dogs we are procuring are too small in most cases and do not have strong enough drive."*

This supports the theory of a double-edged sword. It affects not only availability, but trainability. Scientifically conclusive research identifies the optimum age for a trainable "green dog" as 18 to 24 months. High-end customers willing to pay top dollar decimate that "trainable" commodity inventory. An ominous sign is that the U.S. (military and private canine venues) purchase age of "green dogs" continues to get younger. The average is twelve months. This is one of several warning signs that Europe is on the verge of being incapable of continuing as the primary source for dog procurement. Not only is the decline of available dogs alarming, but there is a second order effect. Once procured, there is no warranty or return policy with European breeders. Once accepted by the U.S., we buy the farm. If a dog fails basic obedience, we assume the liability which is a negative cost benefit. The wash out rate of dogs procured during my tenure was between seventeen to thirty percent. Money wasted. This should raise bells and whistles that the current pool of quality trainable dogs is untenable, and not cost effective. The need for an alternate canine procurement solution is NOW.

I presented the paper written at the International Working Dog Breeding Association (IWDBA) meeting in September 2019, in Stockholm Sweden. The association predicted that available dogs would be exhausted within the next five years. In 2018, the U.S. military had a gap of 270 trained dogs for all services. Add to this the unanticipated cancellations of military and private breeder "buy trips" due to COVID. As the shortfall continues to increase, it has a negative impact on readiness. In essence, the tail is wagging the dog. The K9 community has been aware of this slow decline of trainable dogs for years. Senior military leaders and police administrators—who are holding the purse strings—continue to ignore these warning signs and advice from their canine subject matter experts.

How will this be resolved?

The most significant domestic preference statute is the "Buy American Act." It restricts the delivery of foreign end-products under federal government contracts by granting a price preference advantage to U.S. contractors proposing competing offers. Originally initiated as the "Buy American Act of 1933 (BAA)," it supports procurement of domestic products through a provision of "Made in America." The Federal Trade Commission (FTC) regulates the use of "Made in America" claims and labels. The FTC has published guidance indicating that an item must be "all or virtually all" domestic to be properly labeled as "Made in America." Dogs used and defined as indispensable for national security are a prime example of a product that is manufactured in the U.S. Yet the Executive Agent, ignoring federal legislation, continues procuring dogs from the declining European market.

In 2009, the American Recovery and Reinvestment Act reinvigorated the U.S. economy. Legislative and business solutions to reconstitute effective and efficient dog procurement strategies were enacted through a stimulus package. It encouraged reinvestment in American businesses. This opportunity, presented on a silver platter, should have allowed the DoD to perform a strategic, long-term assessment on overseas canine procurement. Then, capitalizing on the political support for domestic options, it could have introduced the idea of what could have evolved into a national breeding program. Rather the EA remains reliant on the Air Force's failed European procurement and unsanctioned "Foster Program."

As our National Command Authority ignored warnings of the coming attacks on Pearl Harbor and 9/11, the dwindling European breeder dog supply has been documented and realized in the canine community for years. Still the MWD community remains pigeon-holed in their procurement complacency choosing to ignore the inevitable. Both the Recovery Act and Buy American Acts provided opportunities for change. However, rather than look to internally changing paradigms to reliance on a domestic canine enterprise, the DoD continues to rely on European breeders.

The EA turned a blind eye to an economic opportunity that could have ensured canine detection capability to meet National Security readiness. Fortunately, another strategic opportunity to force change may be on the horizon. Senator Richard Blumenthal (D-Conn) has introduced Senate Resolution 4049. It is intended to change DoD canine acquisition practices—and refocus procurement, …through "Buy American" initiatives. We can only hope that the proposed legislation will gain momentum and become law. And perhaps simultaneously provide funding in the National Defense Authorization Act (NDAA) to reverse this downward trend to meet annual trained dog requirements.

In 2017, the American Kennel Club and a consortium of Academics, provided a universal "White Paper." It stated their concern of over-reliance on European-bred dogs—and U.S. inability to meet canine demand. Dr. Cindy Otto from Penn Vet School of Veterinary Medicine presented Congressional testimony, "How Canine Programs Contribute to Homeland Security." In this hearing before the Committee on Homeland Security and Governmental Affairs, on March 3, 2016, she recommended the creation of a National Co-operative Breeding Program. Following a "Buy American" economic strategy. The National Co-operative Program would:

1) remove the reliance on European dogs,
2) save the government millions in transportation costs,
3) provide sufficient high-quality dogs, and
4) eliminate the practice of accepting dogs without a return or warranty policy.

She proposed a unified effort of American stakeholders (law enforcement, military, academia, research and development, etc.) to frame national working dog polices. This is not a new concept. National Canine Cooperative Programs have existed and remain extremely successful within the NATO Military Working Dog Community. For years, NATO nations—Germany, Sweden and Italy—have had programs that provide quality dogs to meet their national security needs. These NATO partners achieved overall success by providing a ninety-six percent rate of trained explosives, narcotics, and mine detection dogs from their internal breeding programs. Their centrally managed breeding programs were crafted to each animal's detection capabilities. These dogs are introduced to obedience and distractors as early as six weeks of being whelped (born). Their habits and personality are observed and recorded. Through character analysis, canine psychologists assess "pup" qualities early and eliminate dogs unable to meet medical and training standards, providing a huge cost benefit in money and time—like the Auburn University breeding design.

In comparison the EA also has a breeding program. A test program never codified as a DoD Program of Record when it came into existence in 1999. Still a test program. The DoD EA breeding program uses pet methodologies. Through a Forster Program for up to eighteen months newly whelped puppies are placed in foster homes of volunteer foster families in the San Antonio area, when available. No indoctrination to performance protocols, no introduction to distractors, no obedience evaluation. Just living as household pets.

The breeding program has failed to meet their production goals since inception. By AF policy they were expected to provide fifteen percent of annual MWD requirements. Never achieved in every year of the "test". Not the fault of the director, a dedicated PhD who has dedicated his soul to the program. Unfunded by the Air Education Training Command, attaining scraps from excess funding from the 341st. No emphasis, but a nice public relations story. Bottom line, the program is ineffective has failed to meet their procurement goals since 1999. Good stewards of the government money?

Department of Defense Directive 5200.31E mandates the DoD to increase acquisitions of domestically whelped MWDs. Congress directed this under section 361 of Public Law 110-417, to the extent possible, while maintaining quality and best value for the U.S. Government. Unfortunately, in 2019, the DoD and AKC were about to enter a joint Memorandum of Agreement with AKC to serve as the catalyst for integration with domestic breeders. That idea never saw the light of day, as the AF refused to change their procurement strategy and maintains their European only procurement policy.

The AF contract regulations required any breeder that wanted to present a dog for evaluation had to pay out of pocket to transport them to Lackland. With no guarantees that a dog would be accepted. The cost alone was not a sound business practice for the breeders. What the Joint Services and AKC proposed was that we establish regional evaluations. We would form evaluation teams, conduct procurement evaluations at designated regional locations across the country, starting twice a year. Removing the incredible financial burden to the breeder. It would have stimulated eager breeders, helping to end the era of total reliance on European only procured dogs.

Inter-Agency Collaborations

Because of the inability to fund R&D and produce the numbers of quality of dogs for the DoD, the JSMWDC also recommended increased partnership with academia. As I've mentioned, in the United States, Auburn University assumed the leadership role in an innovative breeding program that encompasses genetic selection, medical screening, and—like our NATO partners—early introduction of obedience and distractors. Auburn incorporates ongoing veterinary scientific research to improve working dog quality and performance. Their focus is on single-purpose, explosives detection dogs. They've crafted a special capability detection dog used by multiple government agencies. Their triad approach to breeding added a new dimension to canine explosive detection. Through their unique selection and breeding protocols, they transformed canine explosive detection. Both in performance and longevity of their product. They became the primary source for TSA detection dogs…and have provided their product to the U.S. Secret Service, NYPD, and National Security Agency (NSA).

Dr. Cindy Otto, Penn Vet and Army Futures Command, conducted conclusive research, determining canine capability to detect trace amounts of COVID. But how has the DoD capitalized on this data? How will dogs be trained in the event of another pandemic? How long will it take and what agency will take the lead? In her Congressional testimony, a collaborative National Co-Operative would be the central clearing house for all government canine training. Alongside a centrally established and jointly staffed Co-Operative breeding program to meet six objectives:

1) Basic and advanced handler training,

2) Research and development,
3) Strategic development,
4) Congressional oversight,
5) Congressionally approved appropriations,
6) Standardized curriculum.

The end state would be nationally recognized dogs that are fully interoperable. The cost benefit would be in the millions. Obviously, the third element—Congressional oversight—is critical. They must appreciate the program's National Security role. Simultaneously driving the train for federal appropriations is needed to establish and advance its operational, tactical and strategic objectives. The fourth element is the documentation of canine standards. The FBI led that effort. In 2021 national canine standards for explosive detection canines (EDCs) were published by The Academy Standards Board. Explosive Detection Dogs Standard 092. A 10-year initiative, providing the template for training, certification, and utilization of explosive detection canines. Equally important, it established common lexicons, providing definitive clarity to canine and EDC terminology. The was a unified effort brining on Federal, State, and local canine agencies as stakeholders.

The hope is that this guidance will be adopted by first responder canine units, to break out of the current quagmire, with agencies relying on numerous regional training standards and certifications. From an oversight perspective, effectiveness and efficacy would be best served under one standard. Without one standard, the canine community will be unable to affect interoperability.

There was a glimmer of hope that a template for a national breeding program would come to fruition. In 2003 TSA was on the verge of the final stages of implementing a state-of-the-art breeding program. Effectively it would have served as the template for the national breeding collaborative. Fully funded, the trigger was about to be pulled…then overnight it was scrubbed. Funding reallocated to TSA technology. The TSA initiative would have led the way to eliminate reliance on European dogs.

What's the way ahead?

Academia, in particular, Auburn and Penn Vet, remain leaders in canine-breeding research and development for medical and explosive detection dogs. Purdue and Southern Illinois University also serve as catalysts for improved canine physiology and performance. There is also an unsung hero in the breeding and canine training world…a unique group of dedicated individuals who are engaged in training dogs for first responders. Yet they are never given the credit they deserve, "States' Department of Correction Puppy programs." They have been raising and training dogs for the first responder community for years. These programs not only help with inmate rehabilitation, but they provide cost-efficient, extremely well-bred dogs to serve in law enforcement agencies throughout the country. Regardless of inmate status, all are hand selected and espouse the tenets of dedication, devotion, and commitment—and as such, they are contributing to the national security effort. "Judge not yet ye should be judged yourself." This is another hidden resource that could be added on a large scale to the shopping cart of a National Collaborative effort.

We must look at the past as a solution for the future. The AKC WWII Dogs for Defense was a national program that worked. Across the country, pet owners stepped up to support the war effort. Civilians donated twenty thousand dogs to the cause. Think of it! Three programs, linked together in a common goal… reversing our reliance on European dogs. That's where the EA comes up short. There is no strategic vision.

The challenge is to change the institutional and organizational cultures in a procurement process that is neither cost effective, efficient nor workable. Think of it, the opportunity to have a never-ending pool of green and pre trained dogs (obedience), that would reverse the reliance on European Dogs. That's where the EA fails. Simply stated, the EA MWD program is a failed business, because of zero strategic vision. If this were the civilian sector, the Air Force would have gone bankrupt years ago. In a world of demands for national change, a national canine co-operative —and the unification of canine stakeholders is the strategy. If we continue down this path of complacency, and bureaucratic manipulation, canine capability gaps will continue to expand. We will go right back to entering a conflict without the necessary resources, funding, or ability to provide canine force multipliers to the field. Analysis of Joint Service MWD statistics identify the numbers of dog being retired and retirement projections for the next five years. It's a picture of the demand out-pacing production. Will we be able to meet wartime and domestic demand?

Funding

The DoD does not have a Congressional line item of accounting for the canine program. The program is funded as an Air Force School through the Air Education Training Command (AETC). AETC frames the funding requirements for every Air Force Schools, including the 341st. The annual program management objective (POM) for the 341st Training Squadron continues to be based on an evaluation of costs projected from the average of costs from the previous five years productivity. Oddly enough the stakeholders, the 341st nor the Joint Service MWD Committee are consulted or provide input. AETC's MWD cost assessment has not been adjusted to the fair market value for over fifteen years. The average MWD allocation is approximately $4500 - $5000 per dog.

The fair market cost of domestic and European commercially bred "green dogs" (no previous training" is approximately $12000. As the primary procurement focus remains on European Breeders— and competition since 9/11 has grown exponentially, the DoD remains well behind the competition. Its no secret that quality dogs go the highest bidder. Competition from China, Russia, Saudi Arabia—are willing to invest in green. They buy green dogs in bulk for up to $30,000. This bulk purchase practice is draining the pool of available dogs. This is critical MWD strategic intelligence. How has this intelligence been analyzed? Good question.

AETC averaging budget formula almost allowed a catastrophic disaster in 2019. AETC perform their averaging of the 341st Training Squadron productivity. In doing so, they used production figures from 2013 – 2018. In each of those years, the Air Force fell well below their annual production output. Using these negative procurement years, AETC assessed the requirements for the 341st FY 18 third quarter procurement, reducing their budget by $1.3M. Neither the 37th Training Group nor the 341st TRS or the JSMWDC were consulted to confirm this analysis. AETC assumed that the trained dog requirement for the failed production years would automatically reflect production rates for 2019. With changes in acceptable breeds, and the institution of the fair share model, the 341st met trained dog requirements for the first time in over a decade. AETC *assumed* 2019 would also result in failed production numbers. You know what they say about assuming.

The catastrophe was averted only through the immediate identification of this shortfall by the 341st Commander and Director of Operations. Bottom line, the 37th Group had to rob Peter to pay Paul to cross level funds from another line of accounting. This would have never happened if the EA, the JSMWDC and 341st were included in the AETC budget cycle.

Fast forward to the U.S. Northern Command (NORTHCOM) Annual International Chemical Biological Radiological Nuclear (CBRN) Summit 2019. As the DoD MWD PM, I was invited to sit on a panel with Veterinarians to discuss the effects of exposure from toxins, chemical/biological agents on dogs. The invitees were distinguished experts from U.S. governmental agencies, the military, Research and Development partners from the U.S. and Canada…and, to my surprise, representatives from the Office of Management and Budget (OMB). As we presented our panel discussions, the conclusion was unanimous. There was a huge gap in protective equipment for military working dogs.

A senior official of OMB who seemed very interested in attaining more information on the budget shortfalls and gaps in the MWD program approached me. In our discussions, I discovered that OMB was willing to identify this gap as part of the National Defense Authorization Act. This fluke meeting opened the doors and started a process that has since led to the initiation of a Capabilities Based Assessment (CBA). This is a process to attain a DoD MWD Program line item of accounting. A conversation with a passionate OMB husband and wife team jump started the beginning of independence for the DoD canine program!

Failed Business Practices

The current domestic Air Force procurement contract provides no business incentive for private vendors to work with the government. The current contract requires domestic breeders to bring their dogs to the 341st TRS at Lackland AFB to be evaluated for possible selection for the DoD program. They must transport their dogs, at their own expense, to Lackland from across the nation, with no guarantees that any of their dogs will be selected. Historical analysis shows that the acceptance rate is extremely low. In 2018, the DoD accepted one dog from a domestic breeder.

In 2018 – 2020, I attended the American Kennel Club Conference in North Carolina. I was provided time to discuss the DoD program in an open session. Vendors expressed the same concerns that were presented to the EA by the JSMWDC committee. They wanted to do business with the DoD but re-stated their reluctance because the Air Force contract provided no cost incentives. Breeders recommended that the DoD establish a regional selection process. This would be an incentive to present dogs for selection while reducing vendor transportation costs. They have also voiced their concerns that the first time they are aware of the selection criteria is when the dogs are put through their paces at Lackland. The handwriting is on the wall. Modify the contract and effect collaborations to establish regional selection sites across the country. Provide breeders with the selection criteria, allowing them to meet the specific training needs of the DoD. We were singing the same song; however, the Air Force has not evaluated these recommendations leaving the vendors financially unable to provide quality dogs to DoD.

What's the solution?

The Air Force works against itself, protecting their manning, in particular Air Force Colonel (06) positions and now the upgraded DoD PM GS-15 position. The Air Force in fact did initiate their own program review in 2019. They employed my mentor and former DoD MWD Program Manager, to conduct an independent analysis. Bob did a full assessment of the 341st and Air Force Security Forces integrated efforts over the course of three months. Cradle to grave. His final report identified several trends in training that needed to be modified by the 341st. Bob's report also recommended that the U.S. Secret Service tasking cell be moved to DC and placed under the command and control of the DoD PM. Have we heard that before? He also recommended additional manning to support the DoD PM position. When Bob was the DoD PM in the mid-2000s, he was staffed with three Airman: A Master Sergeant who was his deputy and two airman who coordinated the bulk of the Secret Service missions. All worked in one work area. When I assumed the position, I was a force multiplier of one.

Bob presented his program assessment to the Director of Air Force Security Forces. I was privy to the report, Bob and I compared notes, especially on his recommendations to move the USSS tasking cell to DC. But when the briefing occurred, I was not asked for input nor to attend the briefing. While his recommendations for modifications to the 341st training program were accepted, no other recommendations were.

After battling for improvements and strategic vision, and challenging antiquated policies, I was tired fighting tunnel vision leadership. I retired in April 2020. Miraculously my position was readvertised as a GS 15. Many of the issues I identified in my daily battle for program review to implement, and for which I had been labelled as "gone native", were included in the "updated" job description. I was a vocal outsider and had proven experience in every aspect of "real use of canines in a theater of war." But I didn't think blue and my counterparts, all who had led canines in combat, knew that without change, lives would be in jeopardy. Of course, the new GS-15 was a former Air Force Security Forces Colonel, supporting the "no 06 left behind" policy. Canine experience? None.

I refused to join the Air Force institutional mind set. As the DoD MWD PM, I had absolutely nothing to do with the AF. My position was to represent the Joint Services, to prepare canines to meet global challenges. And that's what I did.

Chapter 22 – Capabilities Based Assessment

The Capabilities Based Assessment (CBA) was going to be as important as a blood transfusion. It would provide the ability to be released from indentured servitude under the Air Force Air Education Training Command funding strangle-hold. The DoD MWD program would be recognized as a funded Program of Record. I've mentioned the CBA previously, but this chapter will provide you the layman's version.

Our partners in the Department of Homeland Security were receiving millions through Congressional appropriations. Why? They were not internally sequestered within the DoD bureaucratic chain of command. DHS had direct lines of communication with Congress through the Congressional Committee on Homeland Security. This Congressional support achieved legislation to provide:

- R&D funding,
- TSA Explosive Canine Program received millions for growth and development.
- Guidance to lead Homeland Security partnerships to establish academic breeding collectives.

The DoD desperately needed to pursue its own funding. LTC Jay Coisman, who spearheaded the Joint Service Research and Development Sub-Committee had previously worked on a Defense Health Agency "Capabilities Based Assessment." The CBA is the cradle to grave investigation of program financial solvency. Done by an independent DoD contractor, it also assesses program gaps. In the canine program, the "gap" is that it is unfunded. The goal of the CBA was to perform an internal and external assessment of the program. The program would provide the justification, and present conclusive evidence of the need to have a dedicated Congressional approved DoD line of accounting.

During the 2019 R&D symposium, Booze Allen Hamilton and our DoD Budget team explained the process, costs, and the flow of data through the bureaucracy with the Joint Capabilities Integration and Development System (JCIDS). And the follow-on process going to the Joint Requirements Oversight Council (JROC). Then with information needed to support Defense Acquisitions—and the Planning, Programming, Budget and Execution processes. Don't hide! The wickets are many in the government, especially when it comes to spending taxpayer money. A good thing. The process justifies expenditures and appropriations supporting the National Defense Strategy (NDS), National Military Strategy (NMS) and National Strategy for Homeland Defense. It's just the system.

This is a multi-layer bureaucratic process that needs to be followed, and the cradle to grave review takes approximately a year to complete.

If the CBA failed at any of those decision points, we would remain under the Air Force Education Training Command's medieval line of accounting. R&D would remain non-existent. The committee agreed that acquiring an appropriated line item of accounting should be our number one priority. In close competition with MWD protective equipment. This was going to be an interagency effort. The challenge to kick start the CBA was attaining funding to pay for the assessment. The Defense Health Agency was leaning toward taking the lead, but we also needed the EA buy in.

Fortunately, we were able to get on the calendar of the Security Forces Deputy Director. She was a true visionary, an open-minded senior executive who always supported positive change. The decision briefing got her attention. We provided the facts. The program was stagnant. We were hamstrung and floundering. We were the only DoD Program of Record—with no line of accounting. With multiple medical, operational, and training gaps, the MWD program needed to make major adjustments to sustain the DoD canine enterprise.

The Deputy Director gave the green light and with her concurrence the process moved forward. Finally, we were looking at changing twenty years of business as usual and bringing the program into the 21st Century.

Where is the CBA now? It should be close to completion. My hope is that the intended purpose will leap over bureaucratic hurdles to make the needed improvements to build the MWD program of the future.

Chapter 23 – R&D Inter-Agency Collaborations

Our Federal Agencies: TSA, FBI, DHA are all actively engaged with R&D, and have seen the importance in extending the need to reach out and partner with academia. Academic institutions specializing in veterinary medicine and breeding programs – Auburn, Pennsylvania Vet, Southern Illinois University, Purdue – to name a few, are actively engaged with the Department of Homeland Security conducting performance and longevity studies. While the EA interest in canine R&D is stymied due to lack of funding. Not to say great strides have been made with the creation of the Joint Service MWD Committee Research and Development sub-committee and establishing a DoD Research and Development Forum. But their defined purpose is hampered by the lack of appropriations.

Regardless of House Resolution 302 (PL 115-254), previously discussed, that directed the Department of Homeland Security and Federal Aviation Administration to unify efforts to develop national canine standards, the challenge exists. HR 302 not only focused on bringing together federal agencies, but also working closely with canine stakeholders in assessing and prioritizing research and development projects. A common goal across the canine community. Yet the DoD neither benefited from the joint interoperability with our sister agencies, with the pipeline of Air force funding solely dedicated to procurement and training dogs for the services. R&D is a bastard stepchild. What there was not…Canine Equity.

Where is the Defense Health Agency providing studies on global zoonotic diseases? Where is the collaboration with the Centers for Disease Control and World Health Organizations? And as we languish in zoonotic limbo, diseases like Monkey Pox leptospirosis, Ebola, and anthrax remain threats to both the animal and human population – across the globe.

We have no studies on global environmental impacts on MWD performance. What happens when we drop a dog team into Korea in the middle of winter…how will the dogs adapt, will the dogs scent detection system be impacted by the extreme cold, will we have protective equipment to protect the dog against the extreme cold? What about the heat of the middle east, and the intense humidity in the jungles?

Only one environmental performance study that I'm aware of was conducted by Natick Labs in 2018. In collaboration with the Army MWD Program, a yearlong study to evaluate canine performance in extreme hot weather was performed. The results were disseminated to the services, DoD PM, our NATO partners, and veterinary community. Little or no attention was given. The question lingers: Will different global environments influence canine performance? MWDs are globetrotters and could be tasked to deploy to all corners of the earth, each presenting a different set of environmental issues. Data should have been collected to assess these effects and after each of our nation's wars; WW II, Viet Nam, Operations Iraqi and Enduring Freedom, environmental studies should have been initiated. How will our dogs perform if we deploy to the likes of the Ukraine or the Pacific theater?

Joint Improvised Explosive Detection Defeat Office

This is the history of the establishment of the Joint Improvised Explosive Detection Defeat Office (JIEDDO) in 2006 during my assignment at U.S. Central Command (2006-2012).They were allocated millions of dollars.to conduct research and development. Their mission… to develop systems to jam and prevent remote detonations. The jamming systems emitted a frequency to jam the signal from the explosives device. The majority of JIEDDO systems were vehicular mounted. Yet one counter IED detection system was never included or discussed in JIEDDO's counter IED systems…MWDs. I cut and pasted just a few of these million-dollar systems. Bottom line, dogs were not even a twinkle in JIEDDO's eye. Let's look at the systems that were funded or were on the drawing board to support mounted and dismounted troops:

1. Duke Version 3 Vehicle mounted CREW system: "Duke" is a counter radio-controlled improvised explosive device (RCIED) electronic warfare (CREW) system. Developed to provide U.S. forces critical, life-saving protection against a wide range of threats. It is a field deployable system that was designed to have minimal size, weight and power requirements while providing simple operation and optimal performance to provide force protection against radio-controlled IEDs.

Duke Version 3 Vehicle mounted CREW system.

2. Convoy Jammer System HP 3260H: Modular jamming system intended to protect. surrounding vehicles and personnel against RCIED. Designed for maximum frequency coverage and protection range, the system is used for both civilian and military motorcades. Thor III dismounted CREW system: The purpose of the Thor III dismounted system is to provide the user in the field with a wearable Radio-Controlled Improvised Explosive Device (RCIED) jammer that has been designed to counter an array of frequency diverse threats. The system is designed to jam multiple signal sources that allow it to counter multiple simultaneous threats.

HP 3260 H convoy jammer system installed in a GMC vehicle.

Thor III dismounted system

3. IED Countermeasure Equipment (ICE): In the fall of 2004, the at White Sands Missile Range in New Mexico and New Mexico State University's Physical Science Laboratory developed a jamming system that uses low-power radio frequency energy to block the radio signals that detonate enemy IEDs. The IED Countermeasure Equipment is typically mounted on a vehicle. Used to neutralize IEDs when avoiding, disarming, or destroying them is not practical.

What About Canine

During the open check book years of JIEDDO, only one canine initiative was funded. It was a project, that wasted over a million dollars, without the input from MWD subject matter experts. A scientific belief that dogs handlers could be interchangeable…and that you could put several handlers on one dog. Those of us in the canine community voiced our belief that canine effectiveness and efficiency was achieved by the bond between the handler and the canine. Scientists and researchers are not dog handlers. While they initiated the test protocols, they quickly realized by sheer theoretical analysis that the dog handlers were right, and their scientific analysis was wrong. This was the first and last project done to evaluate MWD effectiveness. That was the beginning and end of JIEDDO's inclusion of research of MWD's as a counter-IED system.

In the spring of 2007, CENTCOM and JIEDDO commenced weekly meetings to discuss funding, strategy, and initiatives to support the counter-explosive effort. I was asked to attend along with my partner in crime, Mr. Reggie Smith. We'd sit there for up to two hours, and not once was canine ever mentioned. Dogs were not even considered an explosive detection system. They were just dogs, and no one blinked twice on the importance and successes these mobile, agile, canine detection systems had demonstrated in Iraq and Afghanistan. JIEDDO visualized that vehicular mounted and man pack systems would be sufficient to provide standoff protection for everyone, including MWD teams. Developing a cone of protection for dogs was "assumed" to be provided under the guise of existing counter-IED systems. The reality, MWD teams performing route clearance and supporting maneuver units were totally vulnerable.

Canine THOR

THOR was a manpack that would emit a protective bubble from remote detonations up to approximately thirty meters. Under false assumptions, commanders believed that MWD teams would be protected from remote detonations by an infantry soldier carrying a THOR electronic counter IED manpack. A tactical error. The THOR manpack was effective to provide a protective bubble up to thirty meters. When dogs pick up odor in the "scent cone," they follow the odor to source which often takes them well outside THOR's protective bubble. A thirty-meter bubble was a good assumption if the MWD remained in the confines of an infantry squad where they were well protected by THOR. MWD teams would venture well outside the protective bubble, unprotected. That was their mission. To date, JIDA has yet to look to the future and capitalize on developing a specific Canine THOR to provide that protective bubble.

Remote Revisited

Protection of MWD's from remote detonations did not come to the forefront again until 2015. A canine R&D company, IK9, had developed a canine electronic jammer prototype. They had reached out to the canine community, and we all circled the wagons eager to receive a demonstration. It was already patented, and ready to purchase off the shelf. From the Army perspective we still have MWD teams serving in the middle east under the command of Army Central (ARCENT). ARCENT confirmed that there was a need for a canine jammer and signed on identifying the jammer as a war time need.

In turn, we went knocking on the door of the Army Rapid Equipment Fielding (REF) Agency. REF is the agency that provides funding to procure equipment identified as an immediate need for deployed forces. REF agreed to sponsor the demonstration for military and law enforcement canine technical experts.

The concept was simple. What looked like a cell phone, weighing approximately a pound, was mounted on the dog's harness. Once activated, the jammer emitted a signal that created a three-to-five-meter electronic protective bubble. For an on-or off-leash search, protection would be provided for the dog. As they say, the proof is in the pudding.

On the day of the demonstration, we mustered at Ft. Belvoir. The demonstration site was an open area field. The format was easy…a linear configuration with boxes placed ten meters apart. Cell phones were chosen as the test item. Half of the observers placed their cell phones on the boxes. The other half kept them so that they could call the phones. The test dog with the jammer mounted on their harness and handler would walk approximately fifteen to twenty meters from the line of boxes. When they started, the observers made calls to the phones on the boxes. If the jammer functioned properly as the dog walked past the boxes, within the three-to-five-meter bubble, the cell phone signal should drop.

We all stood by our respective test boxes as the "jammer team" started their trek. As the dog approached each box one by one the cell phone signals dropped. They ran this test protocol three times, and each time the signals dropped. The jammer was a total success! I was excited that we would be able to provide this data to ARCENT. In turn, I expected ARCENT would make a formal request to REF to have these jammers procured and shipped overseas. Something we'd been waiting more than ten years for. We'd finally be able to overcome this hurdle! I was very much mistaken.

As the MWD Program Manager, I believed that OPMG would be the coordinating agency with Rapid Equipment Fielding Agency (REF). I thought we'd prepare the request to have REF fund and procure the K9 Jammers. But OPMG was an army staff agency, with no command authority. The requirement and justification had to come from the end user, ARCENT.

We needed to coordinate with Army Central (ARCENT) the senior army headquarters to prepare the justification, an Operational Needs Statement (ONS). We worked closely with the ARCENT MWD Program Manager to pull together the justification. Once completed, it was passed up the chain of command for approval by Army Central (ARCENT) Commander. The jammer, like all electronic devices, was a classified piece of equipment. The only negative shortfall was that the jammer was set to jam certain frequencies. Its components-controlled frequency modulation. How could the jammer be disabled if it fell into enemy hands? A technical modification that needed to be reviewed.

But we didn't get that far. To our chagrin ARCENT denied the request. They stated that the infantry THOR manpack provides sufficient stand-off distance to create a protective bubble for MWD teams. However, that decision did not entertain the facts, which were clearly articulated in the needs statement. The denial terminated the project, and it died a less than honorable death. As the MWD community has said repeatedly, the DoD EA has no funds for R&D. Again, no funding and no representation at the big boy table. K9 remains the forgotten stepchild. Until the shit hits the fan, and everyone wants a dog! Hopefully we will not have to go back to the drawing board and start from scratch. The canine jammer was a home run, not only for the military but for our first responders. A proven device to protect our dogs from the deadly effects of explosives! Imagine.

Blue Force Tracker – NYPD Transit Bureau

After 9/11, the concern for the possible introduction of chemical–biological and radiological agents became a high priority. Anthrax had been sent through the U.S. postal service to New York City, Washington DC, and killed a postal worker who had been exposed to anthrax at a postal distribution center in New Jersey. We were in a quandary. How do we detect and protect?

The NYPD remains the only police agency to provide protective equipment that will identify exposure to a radiological threat. The NYPD Transit Bureau Canine unit, in collaboration with Blue Force tracker, procured a number of these radiological monitors. Mounted on the dogs' harnesses, they provide an early warning detection capability. These monitors can detect and alert the officer of exposure to a radiological contaminant in the most highly traveled confined space in the world, the NYC subway system. Through our community of effort and unofficial exchange program, Lt John Pappas and SGT Randy Brenner along with representatives from Blue Force Tracker provided a system demonstration at Ft. Belvoir VA. In the spring of 2016, the NYPD demonstrated the radiological detection system, but also introduced the world to the enhanced capability of a recent modification… the inclusion of GPS.

The radiological monitors were installed with a GPS tracking device, allowing dogs to be tracked and serve as a geospatial platform. GPS monitoring capability would allow MWD teams to perform explosive detection searches and as a mobile mapping platform, simultaneously provide confirmation of clearing search grids. Avoiding redundancy and to use troops to clear the same area.

However, bureaucratic red tape hindered efforts to provide the military with this equipment. Technology is being developed in the private sector, and military system procurement processes are convoluted. There is a litany of commercial equipment that can be adapted for military use, but our "system" hinders and deters technology and equipment improvement from being fielded. This tunnel vision is a detriment to readiness. While the bureaucratic system for research and development remains caught in a quagmire, our dogs remain open and exposed… defenseless.

Case Study – Who Drives the Research Train

Who is the driving force for funding research in the DoD? The Service Secretaries. Each Service, the Army, Navy, Air Force and Marine Corps has a Research and Development Agency. But they are not funded. The services present their priorities of effort and are rolled up in an Order of Merit List (OML). As projects are accepted, the services are assessed the cost for the R&D organization to conduct the project. It's convoluted. What I believed was requesting an R&D project to the Army R&D agency, who has a canine researcher, would have been the way ahead. As we'll see, the process is yet another bureaucratic quagmire.

Case in point, in July 2014 the Army Research and Development Directorate under Dr. Stephen Lee conducted the first ever Inter-Agency Canine R&D conference in North Carolina. Representatives from the military, law enforcement, NATO partners, and Academia attended. This event laid the foundation to share initiatives and form collaborations. The goal was to bring situational awareness and share current and future canine projects. It helped unite the community. It raised our hopes that we would gain support…especially funding. The meeting opened multiple lines of communication with our global canine and R&D partners.

Fast forward two months (Sep 2014). Based on mutual discussions of canine needs Dr. Lee requested the top ten canine research priorities to present to the Assistant Secretary of the Army (ASA). It was going to be a joint Army–Marine Corps effort. The Marines and Army tapped into Program Mangers, Kennel Masters, Trainers, and handlers to identify our top ten priorities. Because of our similar missions our top priorities mirrored each other. We were hoping that our unity would be a factor for attaining funding. We were anticipating light at the end of the tunnel, finally receiving desperately needed research dollars.

Along with Steve Lee, we waited eagerly for our opportunity to brief the Assistant Secretary of the Army. But it was not to be. Our hopes were dashed as Steve returned from presenting our draft proposals with the dismal news. Funding for canine research was not earmarked in the Army fiscal year budget. The only alternative—as I pointed out, is the service come to the table with an open check book. We had hit another brick wall and an R&D standstill. 'One Team, One Fight" remains an appropriations bridge too far. Held hostage…in canine R&D limbo.

Chapter 24 – Canine Intelligence

There is a perception that the dissemination of intelligence had no major relevance in the MWD program. Dogs go through their initial level III training, are assigned to their services…and for the remainder of their career, they train, deploy, and execute their search missions. Handlers, like all members of the military, receive instructions on how to report intelligence in an OCOKA (Observation, Cover and Concealment, Obstacles, Key Terrain, Avenues of Approach) report. However, they may not appreciate how this limited information could play a vital role in troop dispositions, revising intelligence estimates, and effecting intelligence preparation of the battlefield.

When MWD teams identify threats, it should spark not only the OCOKA report, but handlers conducting an MWD assessment of battlefield threats (IEDs, Chem-Bio). Should MWD intelligence be incorporated in battlefield assessments…absolutely. MWD intelligence will impact utilization, numbers of dogs required, and yes, casualty estimates. Normally MWD intelligence is not incorporated.

As the DoD MWD Program Manager, I reached out through my "family," the Combatant Commands, the services EOD communities, law enforcement (FBI Explosive–WMD, Asia Pacific CIED, EUCOM, CENTCOM IED fusion cells), NATO, and EU. I received monthly, quarterly, and annual roll ups of IED activity reports. The DoD had our thumb on the pulse of IED activity worldwide. We analyzed types of IEDs (explosives), methods of employment, components and active terrorist locations. Then, I established a canine intelligence fusion cell at the DoD. A fusion cell of one. Passing information was crucial for having situational awareness to train and prepare for global deployments.

Our partnerships kept us on top of real time incidents. In 2018, there was a terrorist bombing in the Chelsea section of New York City. Within twenty minutes, I received unofficial information about the explosive used and whether it was a lone wolf or possible cell, followed by a threat link analysis. We also received the same information on a suicide-bomber's failed attempt in the NYC subway system some months later. Our network provided critical information that was passed on to the military and our sister law enforcement agencies. It was a unified effort. This information enhanced our vigilance for possible copycat events. It changed the platform of canine information sharing.

Our expanded role, thanks to the Chief of Staff for FBI CIED Joint Program Office, partnered the DoD with the European Union. The DoD received action reports from our European partners who have felt the terrible effects of terrorism. Our EU partners—France, Germany, UK, and Belgium—had all experienced devastating terrorists' attacks using the volatile TATP as their weapon of destruction. Because of our extended outreach we were able to receive terrorist incident reports on the types of explosives, how and if dogs were introduced, and if available, the identity of the terrorist organization. In turn like the game telephone, whatever information I could share, I sent to my canine partners.

Since 2019, we've been included in the EU–FBI–U.S. CIED working group. This provided the DoD with an intelligence threat matrix of each of these European terrorist incidents. I circulated these threat matrixes throughout the canine and EOD team. But the EA displayed little or no interest. One major relevant fact did come to the forefront, which shocked me, but as the world turns, so does the geographic base of terrorism.

Since 2003, we have believed that IED use is centered in the Middle East, primarily in our theaters of war—Iraq, and Afghanistan. A person on the street undoubtedly equates IEDs to that area of the world. Truth be told, IED incidents occur in South, Southeast and Northern Asia. Comparing our Asia-Pacific

Counter IED fusion cell with U.S. CENTCOM and European Command, there has been a significant jihadist shift to the INDO PACOM region (Pacific Rim). There's been no decrease in volume or tenacity. More than forty jihadist splinter organizations operate in the Southeast Pacific and saturate the Philippines.

This threat cannot be ignored. The issue facing the DoD is – will we be prepared to meet the challenges of detecting new types of IED's and masking agents? Will we be able to mitigate environmental challenges, and adversary delivery systems? Do we have the proper canine capabilities to be relevant for possible incursions in this new demographic? Let's capture that for a second. Each instance of terrorism in Europe has been a suicide bomber. What capability is there that can detect suicide bombers - moving targets? The DoD has no capability. Regardless that JIDA's "refugee" comparative analysis exercise at Ft. Eustis provided conclusive evidence the DoD was unable to track a moving IED. The NYPD dogs proved that only a specially trained air scent dog can detect and track a moving target.

Canine intelligence in the Pacific Rim clearly demonstrates that suicide bombers are a credible threat. Has a canine capabilities assessment been conducted by INDO-PACOM? No, yet the threat exists.

A global assessment of capabilities should light the fire that personal borne detection dogs are needed. With thousands of immigrants fleeing their impoverished nations…and border security non-existent, the opportunity for terrorists to enter our nation with the intent to do harm is alarming. History demonstrates not only the type of explosive, but the delivery system. Human. For those of us who entered the military or law enforcement, we all have the 9/11 mindset. It's not if it's going to happen…it's when and where. Will we be prepared to meet and mitigate the threat?

The threat of war with China continues to loom in the Pacific. Canine force multipliers will be a high demand item. But has our Pacific Command (INDO-PACOM)—and the Executive Agent—performed canine intelligence preparation of the battlefield? Will we capitalize on intelligence and use it to our tactical advantage, or will we ignore the opportunities to develop MWD mitigation techniques, tactics and procedures? Will we be led blindly into an area of operation, unable to detect and deter threats? Will dogs go from being force multipliers to force detractors?

Business Intelligence – A Vain Attempt

The Air Force—on behalf of the DoD—did a deep dive and prepared a "Category Intelligence Report." The purpose was to provide actionable business intelligence to identify effective and efficient canine procurement strategies. The report encompassed both the Military and Department of Homeland Security agency working dogs. It used FY-16 – FY-18 Federal Procurement Data System (FPDS) information as a reference. The two-year analysis determined that approximately $85M per year is used for working dog products and services. The DoD is the largest federal program with approximately 2000 dogs in its inventory. The Office of Management and Budget (OMB) maps dogs under a taxonomy protocol. The DoD falls under a sub-category of *Section 3.1 Security Animals and Related Services*. What does that mean for mission success? Little or nothing.

The first anomaly is that the Air Force—as the Executive Agent for the MWD Program—executes procedures and responsibilities for a joint service program. The report focused on how the DoD would serve as the single procurement source for all agencies' working dogs. Centralized management and standardization are the best practices to afford accountability and consolidation of resources.

Who would roll the dice to provide oversight for a federal agency's centralized procurement process? The DoD Executive agent was one deep—and the program was already divided under at least three Air Force entities. Continuity was basically nonexistent. The report, while providing a generic process, was flawed.

The Office of Management and Budget (OMB) prepared a report on government working dogs. Supposedly the report captured the total funds expended for every government agency program. But the report was erroneous. It did not. It didn't capture expenditures for contract working dogs in Afghanistan. It didn't capture the twenty-eight million dollars, or forty-two percent of funds that were used to build kennels. It didn't do an agency deep dive or a cost benefit analysis. And OMB did not include the costs for training special operations dogs at Von Liche and Cobra Kennels. More disturbing the annual sustainment costs to provide health and welfare, veterinary services were not included.

Gaps existed in the OMB report. But the business Government Category Wide Intelligence Report (CIR) stakeholder reinforced existing gaps. A true and honest assessment, with recommendations to fill the gaps and return to a course of joint federal inter-agency continuity. To no one's surprise, these issues had been presented for more than ten years to the Executive Agent.

The gap recommendations were:

1) Review the requirements analysis section that will assist in benchmarking the current state of the government. The bottom line was that the Federal government agencies—DHS/TSA/USSS/FBI/U.S. Marshals canine programs—are all funded and managed separately. The DoD remains under the umbrella of the Air Force restricting procurement funding. In essence, the Air Force is their own detractor for failing to provide competitive fair market pricing.
2) Perform market analysis to help understand industry capabilities/capacity and what is being done differently in the private sector of similar consumers. The Air Force prevailing methodology remains to procure dogs from European breeders, regardless of the Congressional emphasis of Buy American. The seventeen to thirty percent wash out rate of dogs procured from Europe is accepted by the EA as a cost of doing business. And the Air Force Contracting Command failed to perform a procurement review, or private vendor reviews of the prohibitions of doing business with the DoD.
3) Based on comparisons with federal agencies, a team will assess what is being done differently in the private sector. In the category management area, actions should be taken to reduce/eliminate gaps between current and future government practices. In addition, more effective and efficient multi-agency sourcing strategies should be identified and recommended. Recommendation that the intelligence clearly points to the program detractor and leading cause of procurement and production gaps between the private and DoD, the Air Force inability to manage, fund, develop a sound five-year business plan, or desire to relinquish control.

Comparatively there is a parallel to the CIR and Joint Service MWD Committee after action reviews. This CIR could have been an extrapolated version of a cut and paste of the minutes that have consistently been shelved by the Executive agent for almost a decade. The CIR recommendations run parallel with the JMWDC:

1. Establish Centralized DoD Multi-Agency and Government Wide Funding:

a. Establish centralized DoD, multi-agency/government wide research Development, Test & Evaluation funding lines of accounting. Consolidation of effort to effectively and efficiently manage programs to support common working dog missions.
b. Action by the DoD Program Manager – Defense Health Agency: 2019 established the Joint Service MWD R&D subcommittee, with multiple government, and NGO, academic organizations. Centralized consolidation of lines of effort.

2. Establish multi-agency or government-wide breeding programs.

a. Actions by the DoD PM: AKC and a consortium of academics led by Dr. Cindy Otto presented the need for a national canine breeding program. Breeding programs from NATO/EU, and U.S. Academic institutions far outperform the DoD breeding program.

3. Increase Success Rate of Dogs Evaluated for Government contract purchase.

a. Work with industry to increase the success rate of domestic dogs presented to the government for evaluation.
b. Action by the DoD PM: The current failed contract is cost prohibitive, placing the financial burden on perspective vendors who must pay out of pocket expenses to bring dogs to Lackland for evaluation. The Joint Service MWD Committee endorsed solution, to establish regional joint government selection teams. Increase local breeder partnerships and reduce breeder overhead costs.

4. Evaluate Potential to Use DNA Mapping to Predict Dog's Chance of Success.

a. Establish a project with Academia to evaluate the potential to use DNA mapping to predict a dog's chances of becoming a successful working dog prior to government procurement. Provides genetic evaluation on performance and longevity metrics.

DoD PM Actions: None. Auburn, Penn Vet, Purdue University, have exercised this performance assessment, and continue to develop evaluation protocols to measure longevity, and agility. The DoD has not taken action to establish academic partnerships. The DoD remains stagnated in a "test" breeding program that is incapable of executing scientific engineering. Academic breeding programs would provide a high-performance, high drive dog, scientifically evaluated to meet the needs of the services.

5. Prepare Government-wide National Emergency Response Plan (Dog Shortage)

a. Prepare a government-wide plan to respond to national emergencies which would cause an immediate working dog shortage. The plan should address strategies to rapidly increase industry's trained dog capacity and prioritize government working dog missions. Multi-agency agreements should also be established as part of a plan to outline how agencies can share resources to ensure the continued execution of the highest priority U.S. working dog missions.

DoD PM Action: A plan for national mobilization has existed since 1943 when the "Dogs for Defense" program was established in collaboration with the American Kennel Club, providing ten thousand dogs for service during WWII. The DoD has not initiated planning to meet surge requirements or collaborated with AKC to affect a Memorandum of Agreement, enlisting breeders, pet owners to respond to national security or domestic shortages. The DoD is non mission capable to meet expanded requirements. Intelligence dictates a strategic plan to produce traditional, not nontraditional teams.

Tactical Canine Intelligence and the Category Intelligence Report are useful, but only if their content is analyzed and acted upon. Otherwise, intelligence is useless. And when the failure to act on intelligence is realized with the loss of young men and women, we'll have useless Congressional hearings attempting to assign blame. And the Air Force will be directed to prepare more useless reports directing agencies to implement best practices, that fall on deaf ears!

It's amusing that the agency that has virtually ignored the canine program for more than a decade was tasked to pull together the CIR. Of all the participating agencies, DHS, DOJ, DOS, the Air Force is the agency with no program funding oversight—allowing a third party (AETC) and Air Force Contracting Command to fund the program in a vacuum.

Every point captured in the CIR is critical. Unification and standardization have been the mantra and anthem sung by canine stakeholders for over twenty years. Regardless of outreach, research, and man-hours to complete a comprehensive intelligence analysis, unless the reports receive the emphasis to change a broken system and rebuild it from the ground up, with canine masons, carpenters, and artisans the program will continue to spiral. Just good news stories of hero dogs in the nightly news.

Chapter 25 – Strategic Planning

The EA's sole interpretation of their mission is to procure and produce dogs. As the DoD PM, I framed strategic goals and objectives that WERE the primary focus of my job. They aligned with the SECDEF and Joint Chiefs strategic vision. In an address to the Atlantic Council in October 2020, Mr. Esper—the former SECDEF—re-stated the position of the DoD:

> *"To meet the demands of 21st century great power competition, I directed the Pentagon's Office of Policy to develop a first-of-its-kind, comprehensive strategic approach to strengthen alliances and build partnerships."*

Further codified in an article entitled, "Habits of cooperation and interoperability," it clarifies military and political reasons to continue interagency collaborations. My context was that this was the direction for "all" of the DoD. But not for the EA.

Afghanistan became NATO's marquee mission with the U.S. invasion in 2001. It was the first time in history that the alliance invoked Article V, which declares that an attack on one is an attack on all. The NATO-led International Security Assistance Force (ISAF) was ultimately composed of allies from fifty-plus countries, that quickly pulled together under a U.S. Combatant Command, transitioning to a unified NATO command structure.

In the early years of the war, the running joke among U.S. forces, was that ISAF stood for "I saw Americans fight," or "I sunbathed at FOBs" (forward operating bases, which are heavily fortified and largely safe). The underlying critique was that some allied governments used restrictions called "caveats" to prevent their troops from carrying out night missions, for example, or from deploying to certain more violent parts of the country…and, as a result, U.S. and other fighting forces carried a heavier load.

"Still, the cooperation was a growth experience for the alliance," says Ian Lesser, executive director of the German Marshall Fund in Brussels. "These caveats did in some ways hinder the ISAF's ability to operate, but it operated nonetheless, and learned a lot by that in terms of habits of cooperation and interoperability that were tested every day."

At the same time, the experience transformed the militaries of many NATO member nations. In Germany, some ninety thousand troops have deployed to Afghanistan over the years. "There's no German General today who doesn't have military or even fighting experience there," says Markus Kaim, senior fellow at the German Institute for International and Security Affairs in Berlin. The same goes for a generation of soldiers in Italy, Spain, the Netherlands, and Canada and more than fifteen other key coalition partners.

Member forces grew accustomed to collaborating on intelligence sharing and mission planning that made use of some high-tech systems that many nations wouldn't have been exposed to in peacetime, says Anthony Cordesman, defense analyst at the Center for Strategic and International Studies. This in turn, led to a "much better appreciation for allied capabilities."

And it led to an even greater appreciation for allies themselves, including non-NATO partners, many of whom, like Australia and South Korea, took part in the war in Afghanistan.

"If we think about any military engagement of NATO going forward, we'll conceptualize it not as thirty member countries of NATO, but as a loose platform" that includes other organizations and non-NATO partners as well," Dr. Kaim says. "NATO needs partners," he continues, because "NATO is aware that it can't shy away from deep political changes we're seeing."

The NATO 2030 report emphasizes making the bloc a "more political alliance," which means making it a "place where core security concerns of all sorts are discussed," Dr. Lesser says. The Asia-Pacific region, especially China, is a case in point. "It's a recognition that the definition of what bears on Euro-Atlantic security has expanded tremendously."

From this perspective, I've pointed out that the DoD MWD PM was on the verge of establishing a focused training relationship with our INDO-Pacific partners, but the EA made the determination that this relatively new global outreach was not a global concern for the DoD Executive Agent. This was regardless of the handwriting on the wall, ignoring the obvious and the established direction of the SECDEF. Like the pre-Pearl Harbor and 9/11 attitudes, the EA is asleep at the helm.

Our world leaders continue to hammer alliances and partnerships (Mr. Esper, Dr. Lesser), but the Air Force EA remains oblivious of the need to develop this global enterprise. We can expect that when the call comes for MWDs in another global conflict, there will again be an empty page in the EA playbook. We'll be scrambling for MWD teams—and the AF will provide a thousand-mile blank stare when the call comes for surge requirements.

Building partnerships and alliances have been the essential foundation of U.S. Combat Operations since WWI. The U.S. military will never execute contingency operations under one flag. We will attain unilateral approval through the UN or fulfill one of our treaty commitments under a Unified Coalition undertaking. That was proven in operation Iraqi Freedom (OIF) and Operation Enduring Freedom (OEF). That will also be the case when we fulfill Defense Support to Civil Authorities (DSCA) missions requested by state and local authorities. The Air Force EA remains reclusive, or should I say, steadfast in their tunnel vision that their way is the only way.

We learned that lesson early on. We also tackled national exclusion changing the course for unity as part of strategic inclusion. At the onset of the war, many of our coalition partners deployed MWDs, but under their national security policy, were restricted to force protection for their respective military operating bases. If we were going to achieve canine interoperability, we needed to retain all canines under a single operational command. This was achieved with the support of General Petraeus, who provided clarification in updated command guidance, placing all canine forces under the operational direction of the appointed Geographic Combatant Commanders. We added close to five hundred additional explosive detection systems to the force mix. Simple unity of effort and economy of force!

Yet the Air Force has no desire to learn how MWDs are used in combat, or MWD tactics techniques and procedures. What were the lessons learned and how were they applied to meet performance and detection challenges in Iraq and Afghanistan? Is the EA aware of operational capabilities and able to brief the SECDEF on canine operations, as they remain sequestered at Lackland AFB and hidden behind the doors on the fifth floor of the Pentagon? What came to pass was that the AF, without any insight or theater situational awareness in theater, was unprepared to support a canine "SURGE" in Afghanistan. This should have been a signal to the SECDEF about the Air Force's inability to perform Executive Agent duties. The Air Force should have been stripped of their DoD EA appointment. If it happens again, does anyone believe that the Air Force learned from this nightmare and developed a plan to meet global surge resources? Most likely, the Air Force will again divest its responsibility. What was the outcome? Keep reading.

Strategy and interoperability are not a mantra of the EA. Any attempt to achieve the vision of the former SECDEF, Mr. Esper, was blocked. As a result, my directorate had no interest in using my knowledge and experience in Joint operations or exploiting my program management theoretical skills from my Doctor of Public Administration degree. Challenging the status quo and attempting to change failed protocols was unacceptable. What *was* acceptable was bowing down and paying homage to the Air Force chain of command. Acquiescing to business as usual.

We never learn. The EA remains Air Force focused, disengaged from the worldwide stage. It has no interest in strategic interoperability. During my short tenure with the AF, I found there was a total lack of understanding of Joint Service integration. This was restated in the Rand "manning" study. The AFSF does not train to perform joint missions. This is despite lessons learned from twenty years of war codifying that AFSF will be cross leveled to execute non-traditional support for the needs of the theater.

Strategically, the DoD Canine program and our great team on the Joint Service MWD Committee led the way. We didn't let prohibitions and lack of EA support stagnate our mission. With our NATO Military Working Dog Expert Panel, we finalized NATO Military Working Dog (MWD) Capabilities Standard Agreement 2623. This was the burning bush. It was the doctrinal bible for operations under a unified command. It provided common doctrine that all allied canines would operate in a theater of war. Ten years in the making. However, it achieved the SECDEF and NATO goal of interoperability. But what about unified canine operations in the other geographic areas of the globe? How would we project the canine enterprise to support contingencies in the Pacific?

One of my strategic desires was to tackle how we would set the MWD operational footprint in the Pacific should we face conflict with China or on the Korean Peninsula. I discovered that on the fifth floor of the Pentagon, there was an office representing our allied partners from the Pacific, commonly referred to as the Five Eyes. Representatives from the UK, New Zealand, Australia, Canada the U.S. are the interface with the SECDEF. Their mission is… develop interoperable Pacific mutual assistance planning and built collaborations.

What we found was that our Five Eyes partners had not identified MWDs as a priority of effort in any of their war plans. However, they did include MWDs in their regional security exercises, including MWD utilization in Air Base Defense. The first conversation I had quickly opened our lines of communication. Not only would we look to include our MWD teams in future exercises and war planning. We were also going to engage in a strategic goal to assist in joint MWD doctrine.

The NATO STANAG provided that foundational doctrine. Would there be any prohibitions if the Five Eyes adopted the NATO STANAG? We reached out to the NATO Allied Transformation Command (ACT) to inquire on prohibitions and exceptions of the Five Eye Nations to adopt parts or the entire STANAG. We received the response…no restrictions.

One of the most memorable achievements resulting from the great work of the NATO Expert MWD Panel was when the Pacific Five Eye partners made the strategic decision to adopt the NATO STANAG. With a simple cover letter, the Pacific Nations unilaterally adopted the STANAG. In the swipe of a pen the NATO STANAG was now the unified global MWD doctrine. Operational utilization, defined capabilities and canine lines of communication were now standardized across the globe. This team effort achieved global interoperability. No fanfare, no balloons.

The use of military assets supporting civil authorities was put to the test during the pandemic. Active, Guard and Reserve Medical units were called up and deployed to support the domestic effort to fight the war on COVID—the second time the military was called upon to fight a domestic war. The first was the Spanish Flu pandemic in 1918. The military played a vital role in winning the wars on disease.

The military was again mobilized post January 6, 2021. Almost twenty thousand Army National Guardsman from seven states were mobilized to protect the capitol building. And now active troops are being deployed to the Southern Border to support the uncontrolled invasion by illegal aliens. The threat to this nation at this point in our nation's history is not from the Taliban. It's from within. The question must be asked…is restricting military resources from expanded utilization to fight a domestic war on U.S. soil still viable? Is *Posse Comitatus* or the implementation of martial law to protect U.S. citizens still relevant? When men and women take the oath of allegiance to the country we swear "To defend and support the constitution of the United States against all enemies **foreign and domestic**."

The pandemic was a crossroads of how the military is strategically used. But it was not something that we hadn't done before, especially in the MWD community. The pandemic saw the military supporting the tremendous medical effort to combat COVID. The reality is that if the demand for military assets continues the scale of the pandemic, or as was deployed to the Capitol on January 6, the range and scope of the military will need to be addressed. Under Title 10 of the United States Code the military is organized to do one thing, fight our nation's wars. Yet it seems that the war against America seems to be right here in our backyard. Be it a pandemic or domestic threats against the homeland. The question… is the *Posse Comitatus* Act of 1878 still viable in the 21st Century? This does not mean to support political utilization of martial law. Only that we must realize that military needs to be interwoven as a readily available asset to fight our nations battles domestically as well as outside the confines of our borders.

The independent report from the January 6th protests at Capitol Hill identified the need for inter-agency cooperation and use of military working dogs. Specifically identifying military working dogs as a needed counter-explosive force multiplier. Title 10 assets to support federal law enforcement in a domestic mission? Missions on the border, pandemic, supporting the Capitol police. What other domestic missions will the military be tasked to perform outside the sphere of title 10? Our borders compromised, crime out of control. Will it take a Hamas attack on our homeland before we use military resources to protect the country?

Congress enacted House Resolution 302, directing the Department of Homeland Security to provide a study on the efficiencies from attaining explosive detection dogs from partners in academia. The study would be presented to Congress with the goal of Congress considering additional appropriations for breeding collaborations. The DoD had no study and no initiative to change our stone age methodology of procurement.

We remained stagnant, with no funding for similar studies or research and development. I consistently asked myself what the true purpose of the executive agent was. All I ever heard were excuses…"that's a service issue". Useless rhetoric. It was redundant and frustrating. There is no EA strategy for the MWD Program. Throughout this book I've provided multiple lessons learned, lessons that should be uploaded into the strategic data base. For naught. The reality is that all federal sponsored canine programs should come under one national Canine Program Manager and be seated in the National Security Council. The EA remains with their head in the "canine cloud."

Failed Vision

Let's dig further. What have we learned from Operations Iraqi Freedom, Operation Enduring Freedom, and Operation Inherent Resolve? What have we gleaned from the threat posed in the Pacific by China and Korea? Has the Joint Staff or Combatant Commands asked any of these questions?

Canine Intelligence – What are the threats?

1. We know that wherever we deploy, we will continue to face Improvised Explosive Devices. That is the nature of enemy Tactics, Techniques and Procedures.
2. What about subterranean use of dogs? Open-source information has determined that the North Koreans built thousands of tunnels from the North into the South. Will we employ dogs to

mitigate caches of weapons and detect explosives with a high prevalence of losing dogs exposed to chem-bio agents.
3. Is there protective personal equipment for dogs? Is decontamination training a part of a canine detachments mission essential task list? NO!
4. How will we effect resupply for our deployed MWDs? Is there an established pre-positioned equipment site for deploying MWD teams? Not for dogs, but in retrospect, today we have "Human" organizational pre-positioned equipment in Germany, Kuwait, and Korea.
5. Are MWDs included in global war plans? No, they are nonexistent.
6. What is the plan to identify MWD replacements?
7. What are the force flow transportation methods to get dogs into the theater? Are there any "transient" reception centers?
8. Who will determine veterinary requirements (vaccinations, special testing) and has a force ratio been identified between MWD vs Veterinary care?
9. Contract Working Dogs: Is there a plan for utilization of contract working dogs. Has a contract working dog template been duplicated from OIF/OEF. How will contract working dogs be utilized in the Pacific, and other regional contingencies. What will be the cost and mobilization process for contract working dogs. Will we identify placing CWD businesses on "retainer." Doubtful.

We return to square one. Who is the pinnacle to brief these strategic and operational concerns to the SECDEF and Joint Staff? Who works these DoD issues with the Combatant Commands? According to policy, it's the DoD PM and EA. Nonexistent. But when the s**t does hit the fan, the EA will be able to request this information through the multi-dimensional, extremely effective tasker management tool.

Exclusion and Operational Effectiveness

CENTCOM changed the military MWD Program Manager from a military to a contractor position when I departed in 2012. As an officer I was able to develop and direct the theater MWD program and support our forward deployed teams. I had positional authority to help frame the MWD force laydown. As a contractor they have no authority to direct and execute. They gather information, analyze statistics but cannot make decisions on behalf of the government. The position is now a place holder. The mission of the Combatant Commands is strategic, focusing on future operations in their global area of operation.

To ensure canine global strike capabilities, a canine subject matter expert is required on each of the COCOM's and Joint Staff. It is mission critical. Regardless, dogs are tracked as a piece of equipment, assigned a national stock number, and that's not going to change. Simple inventory management. The job is multi-faceted; canine contract management, assessing capabilities, tracking missions, threats, canine logistics, canine intelligence, pre-deployment training, inter-agency and coalition collaborations. A COCOM MWD Program Manager will be instrumental in assessing after action reports, coordinating utilization, and developing tactics, techniques and procedures.

Because of this lax attitude and failure to consider MWD staffing, our MWD teams will be in jeopardy of mismanagement in any contingency or domestic operation. Along with service and Combatant Command MWD Program Managers, we set the MWD institutional and organizational culture. We are also the single point of expertise for Commanders and their staffs.

Do Words Matter

Over the course of the last two decades, the MWD program…even before the acceptable terms of "Woke" and "Cancel Culture" …has been dragged through the proverbial mud. Pressure from canine activists whose love for animals, but total lack of understanding of the military establishment wasted time and effort to ensure that MWDs were not referred to as equipment. We spent months going back and forth responding to Congressional inquiries from PITA and other animal activists demanding that military working dog regulations change all inferences to eliminate dogs from being referred to as equipment.

So, we did. We staffed the changes necessary to eliminate any reference to a dog as a piece of equipment. There is a cultural problem as to how dogs are perceived. In the military establishment they are "dogs". Not systems. A system is made up of multiple components. Systems do not think on their own, they are programmed to perform a system function. So is a dog team. The canine detection system is a multi-component system. The detection system, a four-legged dog, whose components consist of an olfactory detection system, eyes and ears for threat detection, with functioning organs that are the MWDs power plant. No difference from a mechanical system; a tank, plane, ship, or piece of artillery. It's a crew served detection system. Until our leaders across the DoD realize that a dog is a functional system, they will remain misrepresented, with the perceived reality that this detection system is an anomaly.

But let's look at a dog. What is their capability? What does the dog do? Just like a weapon, tank, ship or plane, it is a system. Unique in that it's the only living, breathing system in the DoD. There is a cultural problem for both the DoD and civilian sector and that is the way MWDs are viewed. Are they four-legged furry ass eaters? Or are they pets, or are they a functional operational detection system? As we train pilots, tank, and artillery crews to use a piece of military hardware, so we do with the handler to operate the MWD detection system.

Just like a mechanical system, dogs have "components," maintenance schedules, operational rotations and a shelf life that ensures effectiveness and efficacy. The system is tracked by the Working Dog Management System throughout their life cycle. Handlers perform the basic maintenance, and our veterinarians are responsible for overhauling and performing major repairs. The mind set needs to be changed. Equity in procurement, acquisition and utilization will never be achieved until leaders look at the dog in a systems configuration. Yes, as a piece of equipment.

The facts are clear. A full independent program assessment needs to be performed! The consequences for inaction are—inability to meet demand for efficient, effective, agile military working dogs to provide K9 force multiplier detection "systems" to support national security.

Chapter 26 – Independent Assessment

The Executive Agent, in 2018, requested an independent assessment of the 341st Training Readiness Squadron, to address training protocols and do a generic program review. A former DoD Program Manager, and my mentor, was contracted and conducted the independent analysis. In his final report, one of his recommendations was that the Secret Service Tasking Cell should be moved to Secret Service Headquarters in D.C. He further recommended that the Tasking Cell be under the DoD PM. Codifying the same recommendations that have been presented by the JSMWDC for more than a decade. His analysis confirmed that its current location at Lackland was a continuity detractor.

Enter U.S. NORTHCOM. Northern Command is the senior military headquarters who has responsibility for supporting military operations in the United States, including National Security Special Events. NORTHCOM also had presented recommendations for the realignment of the tasking cell. Their recommendation made sense and complied with the basic concept of unity of effort and economy of assets. They recommended that the two supporting elements—VIPSA and MWD—be consolidated in DC, under the command and control of VIPSA. Consolidation would have ensured administrative support to obtain VISA's, a central point to process travel vouchers, and most importantly, the ability to have dignitary security packages (EOD and MWD) deploy on Presidential support aircraft. How did the Air Force react to these recommendations. Unacceptable. Relinquishing ownership was out of the question. But the Air Force did counter with their own recommendations:

a. The first was to place the cell at Andrews AFB, assigned to the Andrews Security Forces Squadron. That made no sense as the Security forces squadron was stretched to their limit, with responsibility for base law enforcement and supporting security for Air Force one. The question that was even more ridiculous was that not only would the task responsibilities be assumed by the Security Forces Squadron; add to the conversation that the Air Force Security Forces Center would still own the cell.

b. The second recommendation was reassigning the cell to the Air Force Law Enforcement Detachment at Quantico, Virginia. There would be no change if they were assigned to Lackland. Time and distance, VISA applications, the same problems would exist. And who would they be assigned to…guess. Retained by the Security Forces Center at Lackland.

There wasn't even a discussion on the language in the DoD PM's job description that ownership of dignitary protection missions belonged to the DoD PM. There was no consultation with me or the Joint Service MWD Committee. The mission was not the priority…ownership and control were.

The methodology bringing the tasking cell to DC was not Rich Vargus'. It was the recommendations of the JMWDC. My mission was to be their voice. Recommendations always came from the bottom up. Deployment, Secret Service and National Security Special Event After Action Reports intelligently written from the kids who executed MWD "boots on ground" missions drove JSMWDC recommendations. We hid nothing and kept no secrets. Penetrating the Air Force bubble…an ongoing legacy.

National Security Special Event (NSSE) – Funding

Who is the proponent to ensure that the annual budget for US Secret Service National Security Special Events (NSSE)?

Up until 2019 the Secret Service provided a fixed dollar amount covering the annual costs for NSSE reimbursable expenses to the DoD. For MWD NSSE support funding was provided to the Air Force Security Forces Center. For non-election year reimbursements, the amount was a set $400K. This amount was more than adequate. Upon completion of an NSSE MWD teams would submit their vouchers to the Security Forces Center finance technician. She was an Air Force employee, and her job description did not include responsibility for processing NSSE vouchers, determining USSS NSSE budgets or budget reconciliation. Her actual duty position was to manage the Air Force Security Forces travel…not the services. This was an "additional duty."

Annually, and no later than the 15th of October reimbursable funding for the previous year had to be closed out. In the reconciliation process, it was inevitable that service members would still have open vouchers. If they were not submitted and processed by the 15th of October that fiscal year period for reimbursement would be closed. This super star Air Force civilian would end up having to start the telephone tree contacting the service program managers to track down handlers who had outstanding vouchers. A demanding if not insurmountable task. The question… as the DoD MWD program manager with direct responsibility for Secret Service and SECSTATE dignitary protection…why wasn't there a position under the DoD PM to manage this task, a deputy PM or…a travel coordinator?

That was USSS NSSE. Let's not forget that Secretary of State travel was a totally separate reimbursable process. The Air Force GS-11 juggled SECSTATE travel vouchers and reconciliation separately. Those vouchers were processed directly with the travel representative in the Diplomatic Security Service (DSS) office.

Fiscal year 2020 (FY-20) started the election year NSSE cycle and we would be tasked to provide support for the sitting President, Vice President, nominees (if not the sitting President), national Republican and Democratic conventions, all campaign venues, and the inauguration. Each venue normally required one hundred plus dog teams. Would the standard allocation of $400K be sufficient. Analysis of NSSE reimbursements left to the Air Force finance technician, who previously received no input from the DoD PM or JSMWDC.

We changed that and with her amazing expertise, the DoD PM and JSMWDC performed a joint analysis. With input from the services MWD Program Managers, and from the AFSF finance technician, we identified that FY-20 NSSE support was not adequate and would have to be plussed up by $200,000. Requesting the plus up was the DoD MWD PM responsibility. The Security Forces Center had no authority to speak on behalf of the EA. Regardless of the attempts to bully me into not executing my job, with a phone call and email reimbursable expenses for FY-20 was increased to $600,000, sufficient to meet our obligations.

I further clarified with the Secret Service that as the DoD PM that I was the sole point of contact for decisions regarding DoD MWDs, not the Air Force SF Center. Their responsibility…sourcing missions. There was a misconception that NSSE voucher issues were the responsibility of the Tasking cell. They were not voucher specialists. You had an internal gaggle of the tasking cell engaging the Air Force finance technician to resolve travel vouchers. We were going to align the program and re-establish the right lines of command and control, and reverse the hap hazard lack of communications, and provide focus and direction.

The DoD and Joint Services returned to being the driving force for both standard and National Special Security Events (NSSE) as the rightful owner of the program. We established an open line of communication with the Secret Service and Diplomatic Security Service finance team. We changed the paradigm of tracking reimbursements and ensured that the services were executing timely submissions of vouchers.

We looked at best practices to ensure that the DoD was providing financial management as the stewards of government funding. The Secret Service, and Diplomatic Security Service were no longer the enemy placing unreasonable demands on the AFSFC. As part of the Joint Service MWD Committee semi-annual meetings the Secret Service and Diplomatic Security Service finance personnel participated, part of our new family of stakeholders. They were part of a new dawn of cooperation, communications and best practices alleviating years of animosity. These changes improved program management in leaps and bounds.

Chapter 27 – Strategic Lessons Learned

The "Hub" experiences and "lessons learned" were quickly forgotten. But the dogs of war continue to nip at our heels. In 2016 shortly after President Trump was elected, rumors of war with North Korea and China jumped to national forefront. INDO-PACOM started to dust off their contingency plans. Possible conflicts in the Pacific are still simmering. Now with conflicts against our allies in the middle east, there is an ever-growing concern that the U.S. may again be drawn into that conflict.

At the time, I was still working with the Army. Partnered with a first Sergeant Major in the Army MWD career field. We started assessing Army MWD movements and operational utilization in the Pacific. How many dogs and how would they move into the theater? What missions would they be required to perform? What about vets? What about coalition dogs? Would dogs finally be embedded with army units as a deployable force package?

The Joint Chiefs expected that we would quickly deploy forces and conclude operations within 180 days. They said that about the wars in Iraq and Afghanistan, too. Somehow those projections didn't include political blunders that kept us there just a bit longer…years. Under pristine computer simulations it may have worked. When the trigger is pulled all the computer simulations and war plans will go out the window. There will be no mobility or ease of movement for military equipment and personnel. The roads—if they have not been targeted— will be jammed with thousands of refugees. It will be a quagmire. Interstate 95 and the Long Island Expressway on steroids!

But what if our adversaries pull the nuclear trigger? War plans for the most part don't provide a foundation for the execution of ground operations on a dirty battlefield. Under the current climate and recruiting shortfalls, we don't have sufficient personnel to deploy to meet the challenge. What if we don't have war reserves necessary to sustain combat operations because they have been depleted and sent to the Ukraine?

War plans identify the benchmark forces anticipated for fighting a war. They set the stage for the majority of logistical, aviation, naval support necessary. They are all best guestimates. But those of us who have deployed knew that whoever came up with this 180-day projection was smoking dope, looking for a good efficiency report or had been in a magic show performing illusions with smoke and mirrors.

Are our MWD teams prepared to provide and protect troops on the Penn, especially with the two threats that we expect to encounter. Todays' military working dogs are unprepared for chemical and biological warfare. Canines have not been exposed to chemical—biological warfare since WWI. No chem-bio decontamination (decon) training and no protective equipment. Another difference between WWI conditions and now… back then, dogs were provided with gas masks.

Subterranean operations are the other threat. We've known that the North Koreans have spent seventy years building somewhere between 5000 - 20000 tunnels into the South. Each one would have to be cleared or destroyed. A new concept in combat operations is what's referred to as Sub-T …and it is not a task MWDs usually do. While our brethren in Special Operations Command have levied active research to develop decontamination equipment for MWDs, the conventional MWD world lags behind. There are no plans to perform research and development or include decontamination training as mandatory task in the basic handlers' course curriculum. If dogs are exposed to chem-bio, they will become irreplaceable combat casualties.

How would dogs be used as force multipliers once a tunnel had been breached? Do we place dog teams in the unknown without protective equipment to exploit their detection capability? Two problems existed with that. Once breached if the tunnel is contaminated with chem-bio agent…the dog would immediately become a casualty. With no effective decon protocols, what would the disposition of the dog be? Would the dog be field euthanized? Without decon procedures, attempts to evacuate the dog to a Veterinary treatment facility would result in the evacuation vehicle and medical personnel being contaminated.

How will canine assets be prioritized and allocated. Strategically the same mission set will apply to any theater of operation, in this order.

1. Priority to support the Army and Marine maneuver/combat units.
2. Airfield and Naval Port force protection.
3. Forward Operating Base Force protection.
4. Integration of Contract Working Dogs and specialty dogs (cadaver).

The benchmarks for utilization are interchangeable. It will be the canine logistics challenges that will present the challenge. Based on the current threats posed by North Korea and China will INDO-PACOM take heed at the force flow requirements or will they flounder. Key aspects that should already be in the MWD planning phase:

1. How will dogs be transported and supported during transit to and from the theater of operations?
2. If transported through host nations what national veterinary and agricultural restrictions will be imposed?
3. Has the State Department or DoD validated requirements for dogs transiting though Australia, Guam, Okinawa, or the Philippines?
4. Will MWD's be quarantined and if so, what are the quarantine periods?
5. Specific veterinary travel requirements (tests, vaccinations, etc.)

The fly in the ointment is the huge difference of contingency operations between Iraq and Afghanistan and the Pacific. South Korea is "little America East." U.S. citizens working in South Korea, military dependents, and expatriates, would have to be evacuated. Hopefully before the commencement of hostilities. A State Department task, called Noncombatant Evacuation Operations (NEO). NEO will not only be responsible to evacuate U.S. civilians and dependents, but also have the responsibility to evacuate companion animals—pets. How will this be accomplished. In concert with the DoD, military assets will provide the bulk of the NEO transportation, with the emphasis on military air. Adding another layer of coordination, as animals are required to receive a health certificate from a veterinarian to travel. From a U.S veterinarian. In most cases these are only valid for ten days. Who will be the Veterinary/Canine team to plan animal evacuations?

If health certificates and U.S. entry protocols can't be met, do dependents depart and leave their animals behind? Would they be brought back to the U.S. en masse, hoping that sufficient veterinarians from the military, public health service and community can mobilize to support the medical process for their reintegration? Who on the INDO-PACOM/State Department is planning for this?

NEO will probably be conducted simultaneously while forces are being deployed, clogging all major roadways. Let's add to the equation another planning factor. What if simultaneously approximately 500K refugees from the North will be on the march, clogging roads as they flee the combat zone? Under these conditions, how will U.S. citizens be able to move to their designated departure points? And based on the numbers of evacuees, will there be sufficient aircraft to move these non-combatants out of the country? Looking at numbers versus anticipated aircraft needed to accomplish NEO, the demand far outweighs aircraft availability. Winging it when the balloon goes up (when the war starts) is unacceptable. In 2019 the looming threats that MWD's would be exposed to chemical – biological agents, and fentanyl raised the concern of how would we deal with MWD casualties and MWDs killed in action. That was not a presumption, but a reality.

In 2016, the Air Force Director of Security Forces briefed the Army Provost Marshal general on NEO. The Air Force general briefed us that the Air Force would have sufficient aircraft to affect a Korean NEO evacuation. She also assured attendees that she was confident veterinarians would volunteer to support medical screening of the masse of companion animals anticipated to be returning to the U.S. Too bad no one has told the veterinarians in the State of California that they would be mobilized to medically screen returning animals. And if there is not an influx of veterinary volunteers? Will health requirements for travel and quarantine be waived? Someone at the State Department needs to develop that playbook.

Mortuary Affairs

In Iraq and Afghanistan, we lost thirty-two Military Working Dogs and handlers. There was a process in place, for the disposition of our fallen canine heroes. The policy for disposition and return of canine remains was not a veterinary policy, but a U.S. Customs Policy. And it worked. U.S. Central Command Customs Regulation 600-10 provided regulatory guidance for MWD disposition. The policy remained limited to the CENTCOM area of responsibility, never codified in DoD/US Transportation Command policy. In 2019, with the looming possibility of dogs being exposed to chem-bio agents and fentanyl threats, and the looming possibility of deceased animals, policy guidance for MWD mortuary affairs was needed, and quickly.

LTC Jay Coisman, and our CENTCOM Veterinary Staff were the founding fathers of what I believe is still interim policy. Jay was part of that very small strategic group in the Defense Health Agency whose vision was unparalleled. He led the way to develop an interim mortuary affairs format. Using CENTCOM's policy as the template, Jay didn't have to start from scratch. Using the template, he "massaged" the policy, framing an interim Defense Health Agency - DoD policy. One standard process applicable in all global theaters. Written in October 2018, the policy provided guidance for disposition and transportation of deceased "Non-Contaminated" Dogs:

1. Injured dogs would be evacuated from the field (MEDEVACs or moved by vehicle) to a Veterinary Treatment Facility for care.
2. Dogs succumbing to their injuries or illnesses would be located at a Veterinary treatment Facility.
3. Dogs cause of death would be determined by a Necropsy (animal autopsy)
4. Dogs would be packaged in a leak proof bag (Normally a human remains body bag, packed in ice and moved to the cremation facility in Kuwait (Camp Arif Jahn) or approved crematory in the respective area of operations.
5. Dogs' ashes would be placed in a transport urn, with a veterinary certification of cremation and the dogs remains would be returned to the unit or presented to the handler or the handler's family. (Mandatory Customs Border Protection directive).

Note standard interim guidance for "non-contaminated" MWD's. What do you do with a deceased animal whose cause of death was exposure from chemical or biological agents? A good question.

Handlers don't have body bags when they go out on patrol. Dogs are not going to be placed on evacuation vehicles as they will contaminate the vehicle and medical evacuation team. We know that there is no decontamination training or equipment to remove contaminants from dogs. Burying dogs in the ground could possibly contaminate the environment or water tables.

Defense Health Agency worked a miracle in having this interim guidance published in record time. Hopefully this policy has been through the staffing process and in an officially approved DoD policy.

More challenging will be pulling together the MWD, DHA, first responders' stakeholders to develop a national policy to address disposition of "contaminated animals". It's not if it will happen…it's when. Will the community be able to realize the importance of the policy…or kick the can down the road…time will tell.

Not in Your Lane

INDO-PACOM has the lead in the Pacific, working collectively with partner nations to coordinate joint training opportunities and partner exercises. The signature exercise is the annual international maritime exercise, RIMPAC (Rim of the Pacific). Preparedness = Readiness. Along with the multiple joint exchange programs. INDO PACOM works tirelessly to maintain their "strike force" capability. Preparedness and readiness minus one capability that remains in high demand to mitigate explosive and narcotic threats – dogs. Interoperability and integration of MWDs in exercises and exchange programs remains a strategic void.

As the Winds of War subsided in the pacific, MWD integration and interoperability still needed to be pursued. When I assumed the responsibilities of the DoD Program Manager, we had a solid partnership with NATO. NATO MWD doctrine had been written. We were fully engaged with our NATO MWD partners.

In the Pacific interaction remains minimal. Australia and New Zealand had contributed forces and MWDs to the wars in Iraq and Afghanistan and were part of the Pacific Five Eyes partnership. What was needed to establish INDO-PACOM MWD Expert Panel, mirroring the NATO MWD Expert Panel. While my duties, on paper, included inter-agency and international collaborations representing the Executive Agent, I was consistently reminded that my duties should be restricted to the core values of the EA. Regardless, I forged ahead.

Maintaining a continued flow of information with our allies in the Pacific, through our newly established NATO - Five Eye partnership was a dramatic achievement. Through my dear Australian Air Force friend, Alan Grossman I was informed that the Australian Air Force had for years been conducting an annual canine force protection deployment exercise, called Cope North, at Anderson AFB in Guam. Sponsored by Pacific Air Forces (PACAF), Air Force teams from both nations honed their interoperability, and MWD force protection skill sets. The Australian Air Force used this opportunity to run their teams through a deployment cycle. Real world training. The AF kennels in Guam learned how to integrate MWD planning, working together as partners conducting detection exercises. It was training towards a mutual goal, Air Base Force Protection.

From my seat, this needed to be explored. The possibility that we could expand this exercise for inclusion of our joint service INDO-PACOM MWD teams (Army, Navy, Marine Corps). Looking at using this as a catalyst for the inclusion of other Pacific Rim allies; Japan, Philippines, Korea, and India. From a strategic perspective team building initiatives that would resonate MWD interoperability, pre-deployment training and most importantly MWD readiness.

I contacted the PACAF Liaison Officer, a young hard charging Air Force Captain and through his efforts we were able to acquire approval for the inclusion of the other services MWD teams into Cope North. Unfortunately, when I proposed this joint initiative to my chain of command, it was immediately, in no uncertain terms killed. I was overstepping my duties and responsibilities. The EA does not engage in joint exercises. If the services wanted to participate, they could reach out. The EA is not involved with the global theater initiatives, we just procure and train dogs for the services. This was outside my lane, I was not following the elusive core tasks accepting a Pearl Harbor or 9/11 complacent attitude, dramatically affecting MWD global readiness. Cope North remains a PACAF internal exercise, with no outside pacific rim nation participation.

I never gave up the battle and continued to communicate with our pacific partners. And that paid off. In early 2019 the New Zealand MWD Program Manager contacted me. New Zealand was sponsoring a Pacific Law Enforcement Working Dog Conference. Allied Law Enforcement and Military leaders from the region to include India, and the Philippines would be attending. From a strategic perspective the opportunity we were waiting for. Starting the process to cultivate future collaborations, and hopefully aligning mutual training and exercise opportunities. It was hopefully the forward momentum we desperately needed in the region.

I received a personal invitation from the Director of the New Zealand Defense Forces, but again this was outside my realm of responsibility. Not in my lane. I was maintaining my never give up royal pain in the ass strategic team focus. I was the pain that wouldn't stand down. Regardless that the DoD was an essential collaborator, and I was specifically requested by name, I was denied the opportunity to to execute my mission as the sole DoD MWD representative.

But we were able to get representatives from the Marine Corps and Army MWD programs to attend. Along with several NATO partners they established lines of communication that focused on identifying the methodology for future canine exercises, training opportunities, and information sharing. That jump-started the flow of information, but who was going to be the lead at INDO-PACOM? The non-existent MWD Program Manager on the Indo-PACOM staff? Pacific partners continue to engage, but even today the amount of effort being placed on MWD utilization and integration to support the pacific at the INDO-PACOM level is non-existent.

So, who has the responsibility for the INDO-PACOM MWD program? Supposedly INDO PACOM delegated the responsibility to the Pacific Air Forces (PACAF). OK, so when the DoD inquired to the PACAF MWD Program Manager, an Air Force Master Sergeant requesting information of MWD force structure in the Pacific theater …there was a very long pause. The response from their program manager…I only manage Air Force Dogs. No fault of this well-versed Air Force MWD Program Manager. He informed me of the numbers of dogs in the Air Force MWD force structure. He was spot on; he knew the Air Force requirements.

Maybe the U.S. Army Pacific MWD Program Manager has that breakdown. Contacting the Army MWD Program Manager, I already knew the answer, it was non-existent. The run around was ludicrous and not the fault of the services. It's an INDO-PACOM staff responsibility and with no MWD Program Manager…no one tracks INDO-PACOM MWD requirements.

The thought of the immense challenges we faced ramping up the CENTCOM MWD program sent chills through my body. God would the Pacific be a repeat of the travesty that first occurred at the beginning of OIF and OEF. Would we have dogs getting lost in the sauce for up to forty-five days. But Jesus, we had the template for success, the simplicity of common sense. Cut and paste the CENTCOM MWD policy, tailor it to meet the needs of the INDO-PACOM (Hubs, pre-deployment sites, Vet support, canine resupply—prepositioned canine equipment). When and if it happens, everyone will be scrambling to figure out who's in charge. It will take months of starting from scratch and experiencing the same problems we did initially in CENCTOM. And lives will be in jeopardy.

Back to OZ

Strategically the probability that the MWD Transit Hub would be established in Australia is ninety percent. Looking at the same facilities and transportation force flow that we had at the Al Salem MWD Hub in Kuwait, Australia provided the same template for a Pacific MWD Hub. It was designed to be interoperable. However, there was a glitch. The Australian Department of Agriculture's animal entry required a ten-day mandatory quarantine at their agriculture facility in Sydney, *no exceptions*.

The requirements: All dogs coming into Australia require an import permit. Dogs may only be directly imported to Australia from group two or three approved countries. All dogs coming to Australia are required to arrive in Melbourne and undergo a minimum of ten days post entry quarantine at our government facility near Melbourne (Mickleham). The USA is a group three approved country for the purpose of exporting a dog to Australia. Australia has seven diseases that are biosecurity concerns.

Once dogs completed quarantine they could be moved to the pre-deployment/transit hubs, which we believe will be Perth (by air - four hours, by ground - twelve). Completing final processing for onward movement to points east. An essential part of MWD strategic planning. We had a planning foundation.

Another strategic initiative ignored by the EA and INDO PACOM. The pattern of disregarding recommendations, denying access to the EA wore on me for close to eighteen months before I got so sick and tired of the bureaucratic BS, that I decided to retire. But I never wavered, I never allowed myself to compromise, I followed the tenets of General Powell's Leadership treatise.

"I have not allowed myself to be coerced to provide very, very cheap solutions that look neat but won't accomplish the intended purpose."

Chapter 28 – Forgotten Infrastructure

If you've seen the Parthenon in Greece and the Roman Ruins, you'd be assessing the current situation of forty percent of our DoD kennels, being held together by band-aid fixes and bailing wire. Why? Because canine facilities are the bastard stepchild when it comes to improvements and construction on our military bases.

The DoD does what they do best, passes the buck back down to the services to fix their own kennels. Historically over the course of the past forty years, kennels have been designated based on facility availability by an installation commander. In many cases kennels were left over buildings from the 1950/60s, converted into kennels. Once assigned, an Army veterinarian must perform a suitability evaluation. The kennels must meet a specific design and sanitary standards approved. What isn't in the evaluation is the condition of the building or a life cycle assessment. Once approved by the vet MWDs can move into their homes. The responsibility to ensure the safety, health, and welfare of MWDs is the veterinarians, kennel masters and handlers.

In June 2013 the House Armed Services Committee introduced a bill directing the DoD to provide how they would maintain their canine infrastructure. H.R. 1960 House version of the FY 14 Defense Authorization Bill, HRPT 113-102, House Armed Services Committee Report, 20 June 2013: Requires OSD (AT&L) to provide a report to Congress outlining how DoD intends to maintain both the capability and the infrastructure required to support canines as Stand-off Detection of Explosives and Explosives Precursors (SDE2P)

The report to Congress would provide the following information:

1. describe how the DoD intends to maintain the capability and infrastructure required to support canines as stand-off detection of explosives and explosive precursors.
2. specify the appropriate office to oversee the acquisition process, research and development, technology advancement, testing and evaluation, and production and procurement with respect to canines as stand-off detection of explosives and explosive precursors.
3. specify the plan to sustain and enhance the partnerships and relationships of the Department of Defense with service laboratories, private sector companies, and academic institutions to ensure that the latest data and information regarding canine capabilities are distributed throughout the Department and other Federal agencies that could benefit from such information.
4. specify any technologies capable of replacing the canine as a stand-off detection capability during the next two years.

The same issues that have been presented to the Executive Agent for years!! If a report was generated to Congress, it's nowhere to be found in the Executive Agent Archives. If this is an enduring requirement, why hasn't the DoD complied? What has happened and described in preceding paragraphs is a horrendous failure to maintain canine infrastructure. The end state kennels across the DoD are on the verge of being condemned. These same reporting criteria were originally presented in the initial Capabilities Based Assessment evaluation checklist. Removed by the EA. The EA wants nothing to do with anything other than fumbling with procurement and training dogs for the services.

Citizens should be raising hell about the conditions of our military kennels. How was the EA able to avoid this mandated NDAA oversight? Many of the kennels are unsafe and as presented in a previous chapter need major repairs or be torn down and rebuilt. MWD facilities, like all tenant units, compete for installation funding. The installation Directorate of Public Works prioritizes maintenance projects. Unfortunately, when the project is assessed to be outside their funding limitations the units fall into one of two categories. Major Military Repairs are projects that are under $1M, and that funding is controlled and prioritized by the installation commander. New Military Construction projects exceed $1M and I believe that cap has been increased to $2M. New Construction money is not controlled by the installation and those projects are performed on a five-year cycle. Funding approvals occur at the service headquarters.

Each year the installation facilities review board evaluates requests for major repairs and does a "rack and stack", prioritizing projects on an Order of Merit List (OML). The normal course of kennel improvements—the kennels become a top priority, and every year at the installation midyear review kennels go to the bottom of the barrel. The higher priority should be for troop welfare, realignment of units, equipment storage facilities are mission essential and are given top priority. Kennels continue to scramble to keep their heads above water, performing ad hoc fixes, and Army Veterinarians pushing the envelope allowing exceptions to keep kennels open. If installation kennels were condemned, where would the MWD's be reassigned?

There are no funds allocated to the respective MWD programs for anything other than operating maintenance funds (food, replacement canine equipment, computers) basic canine operating necessities. Who should control a separate line item of repairs and construction funding…the EA. Funds should be fenced (untouchable) and controlled and managed by the DoD PM and JSMWDC. But like a Teflon frying pan the EA wants nothing to do with "Joint Responsibility". When I directed CENTCOM canines, I was the sole owner, and my responsibility.

When I was the DoD MWD Program Manager, I was notified by the commander of the DoD Holland Veterinary Hospital that he was filing an Inspector General complaint. The reason was the continued inability of the 341st to maintain the training kennels. Dogs were being brought in with mange because of a lack of caretaker staff. The kennels bases were cracking, filled with feces, and the most astounding issues is that the Air Force Chain of Command at Lackland, the 37th SF Group had been aware of the deteriorating conditions for years and did absolutely nothing to initiate major repairs and hire caretakers.

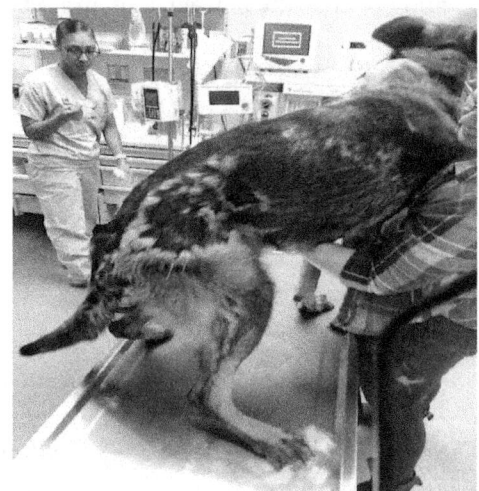

Lackland AFB Training Dog

One of the systematic problems that was created by the air force was their own internal hiring process. Caretakers were graded as GS-4s. Receiving minimum wage, their duties were to sweep out and wash down the kennels. They would walk the animals daily as part of their conditioning. They would provide daily activities report to the Commander 341st, providing their summary of any concerns, such as medical issues or kennel conditions.

But the application process was long and arduous. Caretakers submitted their applications through the Air Force personnel center. You would think that the process of hiring entry level personnel would be quick. No, the Air Force was their own worst enemy. The processing time from submission to appointment could take between four to six months. By that time a large majority would lose interest and the candidate pool was never able to be filled. As the 341st is under the control of the Air Education Training Command, not a a Joint Service Course of Instruction, the 341st was prohibited from performing direct hires. Systematic bureaucratic stupidity, and the typical Air Force way.

I never faulted the Commander or support team at the 341st, they were stepchildren, superheroes who worked miracles with what they had. If this were a true Joint Service Course, they would have their own funding to effect repairs and hire. The 341st fixed what they could internally and weren't going to ignore the situation. The commander took charge and provided an intermediate fix. He ordered instructors and his staff to come in during their off-duty time, to perform housekeeping duties, double duty for what should have been performed by a GS-4 caretaker. Above and beyond all else, take care of the dogs! An easy fix that would have immediately alleviated the backlog in hiring…hire a contractor. Avoids the delays in hiring, no overhead, fixed firm price, no benefits. Simple and efficient.

Not the only issue with training at Lackland. Part of the 341st Basic Handler Course requirements was to perform building searches. The teams are trained to navigate narrow corridors, stairs, and methodically search furniture, which includes emanation of odor from draws, and high searches on high boy closets. The 341st utilized three 1950 Korean-era barracks buildings. They used them for years with zero funds for maintenance. Everyone just watched them die a slow death. Deteriorating to such a state of disrepair that holes in floors and the rotted stair planks had become a safety hazard. In 2019 all three buildings were condemned. That left no buildings to perform building searches.

Temporarily it was recommended that the 341st use old Wilford Hall Hospital that was scheduled to be torn down. That recommendation would have sufficed if the existing structure had not also been condemned. The solution from the chain of command…solicit tenant units and see if their buildings could be used to run building searches. You can't make this up! So, the call went out openly soliciting office space. The course would be modified, and training would be conducted "off duty" after 4:30 pm. This became a handshake deal. The temporary fix forgot that dogs are not house broken. Hmm, what if dogs destroyed property, or worse, dogs urinating or defecating in the "borrowed" search area. What was the way ahead?

The recommended solution was shared facility utilization partnering with the TSA. The TSA Campus Plan a series of joint use buildings, training labs, that would be shared between the TSA and DoD. The problem, the groundbreaking for their construction was still months off. The first building was not going to be a training lab, but an administrative facility. The gap remained, with the 341st struggling to find adequate space to train on building searches. What was not discussed is that there would be a cost share. While the costs of construction would be absorbed by the TSA. The 341st would be responsible for annual operating maintenance costs, costing thousands that the Air Force was not budgeting for. The belief that the TSA campus plan was the solution was not the case. Nothing is free.

The icing on the cake was the report received from Holland Veterinary Hospital on the conditions of the Navy Kennel in Lemore, CA. The report and pictures were disturbing. The response from the Air

Force was horrific. The pictures were horrific and sickening. Immediately I prepared a serious incident report, forwarding it, with pictures up my chain. To my amazement, the Lt. Col., who was my supervisor, informed me that it's not going forward to the EA, it's a service problem. He was serious, I wasn't angry. I was infuriated! "Not my problem!" I was not surprised when I went to the Division Chief, an Air Force 06, who voiced the same sullen complacent response.

Inadequate kennel sanitation (Lemore Kennels)

We don't want to bother the EA; well he was going to be bothered! This was an EA issue, not a service only responsibility. The Office of the Secretary of Defense P&R has the responsibility for DoD Canine Policy oversight. Unhesitatingly I called my contacts in OSD and forwarded the pictures with the IG investigation. Intentionally violating and bypassing my chain of command. OSD was distressed, and shortly thereafter the information was passed up to the P&R SES level II, a three-star equivalent. My understanding was that the phone call with the EA and the SES was not pleasant. Questioning the EA and my division chief on why the conditions had deteriorated so badly, inviting them for a chat, directing an action plan. I did what I had to do, and would do it again, chain of command be damned. Yes, my dear friends you'll see the pattern. I had gone "Rogue" …again. To me a badge of honor. The highest responsibility we have as leaders is being entrusted to take care of our troops, two- or four - legged.

What happened to the kennels after this encounter? I don't know. But what I do know is that the gap assessment to include funding for major repairs and new military construction that was originally included in the Capabilities Based Assessment was conveniently removed. The real problem…the champion of the program is still warming up in the bull pen.

Chapter 29 – No Seat at the Table

In the beginning of this book, I outlined how the DoD has no budget or driving the train on Research and Development. While the services do a fantastic job of training dogs to support their integrated global missions, the EA has no representation in the Halls of Congress, and we have no voice in the House Appropriations or Armed Services Committees. Canine is low on the totem pole. The DoD MWD program is buried in multiple layers of Air Force General Officers who have no clue about canine capability and are charged with protecting Air Force big ticket items. As a result, the MWD detection system program languishes. The only time they demonstrate concern about War Dogs? When they're called on the carpet by Congress or receive negative media attention.

Yet while the DoD sits idly on the fifth floor of the Pentagon, the Department of Homeland Security is glued to their Congressional Committee and since 2001 has received close to a billion dollars for improvements to their program. The new state of the Art TSA canine facility at Lackland AFB, was sourced by a major expansion of appropriations following 9/11. DHS receives millions for their canine Scientific Research Programs, and funding to partner with academia. Increased funding for the DoD program…negligible to none.

Members of the Veterinary Community also have a champion in the Congressional Veterinary Caucus. The caucus is a non-partisan group that advises members of Congress on the veterinary health of the nation. It reaches out to the private and academic sectors, and all the Veterinary schools that are performing genetic research. However, the Air Force remains reactive, responding only to congressional inquiries, avoiding strategic and operational issues.

Without a DoD canine champion in Congress, the next global contingency will result in instantaneous gaps in providing the numbers of trained dogs to support contingency operations. And history will repeat itself. The DoD owns the majority of all federal government dogs. We should be represented by a National Canine Coordinator at DoD and on the Joint Staff. The position would have a finger on the pulse of canine readiness. It would consolidate all stakeholders in support of canine research and development, while assessing threats and new capabilities. And that position would lead the national effort to pursue appropriate legislation to guide, advise and in many cases run interference and protect the sanctity of the program.

However, if the DoD continues its' existing path, we can only hope and pray that complacency will not result in kids dying needlessly on foreign battlefields and here in the Homeland.

Annual Report

I briefly covered the Annual Report to Congress. This will provide you with a better insight of how the Air Force uses their magic wand to pull dogs out of their hat. In Fiscal Year 2017, the National Defense Authorization Bill re-introduced the Department of Defense Annual Report on Military Working Dogs. The report represented Executive Agent statistics on procurement and the production of dogs from overseas and domestic sources. The report was and remains the Air Force Executive Agent scorecard. And the Air Force makes sure that numbers say exactly what the Air Force wants it to say. The report provides a broad whitewash, painting the picture with limited data. Nowhere do they identify the gaps or the Executive Agent's inability to meet and sustain the DoD Canine program for more than a decade.

The report is presented to the House and Senate Armed Services Committees, all nice and neat, with the numbers aligned and totaled up correctly. But these numbers don't tell the story. In this book, you've seen Kennel conditions in the DoD. You've seen data showing that since 2013…with a reprieve in 2019…the DoD failed to meet the services' Trained Dog Requirements. You've been appraised of the EA's failure to meet combatant commanders MWD surge demands. And you've seen the lack of research and development to combat future threats. The breeding program is captured in the report but to the layman, the numbers presented hide the lack of funding, lack of foster families, washout rates from a failed "non-existent" program. The breeding program is NOT a recognized program of record! It has remained a TEST program since it was established in 1999.

The Air Force intentionally shortchanges the information contained in the report, excluding their dirty laundry, sweeping Air Force training and funding gaps under the rug. Rather the AF kicks the ball down the road and allows the conditions and readiness of our canine detection systems to fall further and further into the abyss. When needed to meet increased demands for MWDs, the Air Force will, as they always have, come up with another excuse, pass the buck off to the services, blaming them for production failures.

The annual reports of 2018 and 2019 added information of MWD adoptions to former non-traditional handlers, and euthanized MWDs. While inclusive of this new category, the report intentionally avoided bringing up program gaps. The EA continues to be unwilling to bring to the table existing program shortfalls, identifying the additional appropriations desperately needed to improve canine readiness. Rather keeping the public in the dark with wonderful TV stories of dogs going through an obstacle course or taking down a handler in a red man suit. However, they made sure the cameras didn't pan across to the training kennels. Smoke and mirrors. These are the EA's acceptable "best" practices.

Somewhere in my military and doctoral training, I learned that as a good steward of government funding and as a leader, to move forward, you need to identify and fix your current problems. Gather your stakeholders, identify the issue, allocate the necessary resources, and complete the task. Simultaneously, continue with your long-range projections, identify your timelines for corrective action.

Conveniently, the entire AFSF continued to lift and sweep canine issues under their blue rug. Do not raise SECDEF or Congressional ire. Just let the blister fester until it explodes. The moral of the story, when and where in the world will that blister burst? The result, a nicely written report, properly formatted, with all the numbers matching. So, bewitching and mesmerizing that no one from the SASC or HASC blinks and eye, no questions, the mission to bamboozle Congress achieved.

Chapter 30 – Recognition and Tributes

The Canine Lone Ranger

When I was magically thrust into the fledgling CENTCOM MWD program, I realized that our logistical chain was almost non-existent. There were few dedicated canine lines of communication. Dogs and their handlers deployed as individual teams, except for the Marine Corps. The other services were not embedded or attached to a supporting unit. This situation hindered our ability to support our four-legged heroes.

Sometime in 2006, I received a call, from a canine angel…Ron Aiello. Ron was a godsend. The President and CEO of the U.S. War Dogs Association since 2002, he established "OPERATION MILITARY CARE K-9." They mobilized and were sending care packages to our U.S. Military Working dog teams in the middle east. Ron was a Marine Corps Scout Handler in Viet Nam. He knew the challenges our dogs faced. Far more formidable enemies than our two-legged adversaries. A vicious climate and rugged terrain temperatures of over 130 degrees. The rocks and hot sand burned their paws. The blowing sand stung their eyes. Knowing this, Ron sent practical items to our U.S. Military Working Dogs, things like "Doogles" to protect their eyes, cooling vests to protect them from the high temperature and heat stroke, and boots to protect their paws.

Ron not only provided these "doogles" but whatever else our handlers needed—doggie toys, leashes, bowls, dog food, medicines, vari-kennels. He and his great team still…to this day…support our deployed MWD teams. Even though I was apprehensive at what Ron was doing, his personality and drive is infectious. I embraced his efforts with great appreciation because…like the rest of us…he just wanted to take good care of our MWD teams.

Ron's concern for U.S. War Dogs comes from a time when MWDs were considered expendable equipment. Ron's members were all former Viet Nam MWD Handlers. Army, Marine, Air Force, all Scout or Sentry dog handlers in Viet Nam. Their dogs alerted to multiple threats including punji pits, tunnels, and enemy soldiers. In their time, they saved countless lives. Tightly bonded to these animals, these handlers had to "drop leash" and return to "the World." And sadly, they were forced to leave their canines behind. The concern and policy at the time was that dogs might bring home zoonotic diseases picked up in the jungles of Vietnam. To the bean counters in the Pentagon, the risk was not worth the cost to medically screen these heroes and bring them home.

Ron and U.S. War Dogs were parallel partners in the logistical battle to ensure MWD equipment got to theater. He was the lone Ranger, quietly taking care of our teams, with no fanfare and no publicity. Their mission was, "For the Love of the Dog." The CENTCOM MWD Program relied on Ron, whose members donated thousands of dollars to support our forward deployed troops.

But Ron's work didn't stop there. In a nutshell, when MWDs have been evaluated and determined that their service life is complete and if they meet the eligibility criteria, these hero dogs can be adopted. When this happens, the government's responsibility to maintain the dog terminates. Handlers and the government execute a hold harmless agreement. The person adopting the dog agrees to assume all financial responsibility…and all pre-existing medical conditions. Once that cord is cut, the onus falls on the adopting party. In most cases, this is the MWD's handler. Unfortunately, pre-existing conditions in retiring war dogs…often require lifetime medical treatment. For a young handler wanting to bring his or her canine partner home, these costs could be astronomical.

Ron and U.S. War Dogs refused to accept breaking up a retiring MWD and its human partner because of potential future costs assessed on an adoptee. Ron and all his team would have given their right arms to have had the opportunity to adopt and love their dogs for the remainder of their lives. So, through an endowment, for the past twenty-five years. War Dogs provides free medical care and routine medications for retired MWDs living in a well-deserved loving home.

U.S. War Dogs set the gold standard in one of the most integral behind the scenes partnerships, supporting our canine missions during and post service. Ron has honored war dogs in his home state of New Jersey at the Holmdel MWD Monument, dedicated in 2006. A Lone Ranger, unknown to Command Leadership, but known to the MWD community, a godsend to our MWD handlers and their dogs. Unsung hero…Ron Aiello…Angel on our MWDs shoulders.

Chapter 31 – Exemplary Infrastructure – Army Style

The typical response by the Air Force for everything was pass the buck to the services. Particularly true as we faced the knowledge that kennels had long outlived their construction life cycle. Kennels in comparison to family housing, barracks, and training areas, remain low hanging fruit. That is why the DoD needs to centrally manage all categories for the program and control the destiny that will project an effective and efficient DoD MWD enterprise. It will also take looking outside the box, with innovative ideas. This one, came from an institution of leaders on the Hudson River.

In 2015, OPMG welcomed "Cas", a retired Army MP. She was a fireball, from New York City, and we clicked immediately. We both had distinct New York City demeanors and accents…and our inbred sarcastic sense of humor made us both feel like we were back in the city. As the kennel dilemma continued to fester, Cas, a graduate of the U.S. Military Academy informed me that I may be interested in coordinating with the Military Academy Engineering Department. She let me know that all Cadets are mandated to complete a Senior Project, the projects aligned to an Army mission.

Once Cas laid the groundwork with the Engineering Department, we set up a date and time to meet. Originally from New York, my wife and I still had a home only a few minutes outside of West Point. More importantly, I had the opportunity to have "real" NY Pizza, bagels, and an egg bacon and cheese sandwich…heaven!

The Engineering Department is just south of the parade field, where so many of our nation's military leaders marched across the plain and into history. My meeting was with the Department Chair, their PhD Deputy, Cadet Advisor and three Cadets who would become "Firsties" (Seniors) after their summer training. To be honest, I was overwhelmed by everyone's enthusiasm and support for this project. After introductions, we got right down to identifying the problem and creating the mission statement:

Develop standardized kennels for the Army (DoD): MWD Small (Nine Kennel Runs), MWD Medium (Eighteen Kennel Runs), MWD Large (Thirty-six Kennel Runs). Analyze structural requirements based on the following environmental/demographic considerations: Tropical Environments, Arid Areas, Extreme Cold, Temperate. Identifying sources of cost efficiencies through energy saving systems.

They would identify construction, material costs, and efficiencies for a projected life cycle. All designs would be based on federal and local building codes. The project included Cadets visiting military kennels in the respective demographics, performing site surveys, gathering data, and interviewing Kennel Masters/Trainers/Handlers.

After the project briefing, the Advisor and Cadets discussed specific action items that would need to be completed, in a language that was totally foreign to me. My job was to facilitate. The baton was now passed, the Army and the DoD's aging kennels depended on the dedication of this team of future Army leaders.

These cadets were amazing. They did their mission analysis…and within a month, they had planned and coordinated visits to Army Kennels across the Nation and Alaska. With their advisors, they started developing templates for kennel designs in coordination with the Army Veterinarian at West Point, who rounded out the project team. It was a universal Army effort. The next phase was the site visits which would encompass their Cadet summer training. In essence, this project generated the opportunity to execute a mission needed to support the Army.

This experience was not a college final. This project was preparing them for assignments with the Army Corp of Engineers. They were challenged to build—from the ground up—sustainable, cost effective, facilities. Kennels also required these cadets to become familiar with the MWD team mission. Their MWD education was an ongoing process of reviewing training documents, after action reports, and MWD physiology. Their real education came from MWD experts when they performed site visits. Without hesitation, we placed our trust and confidence in the abilities of these cadets who demonstrated professional qualities of leadership, teamwork and the highest ideals of duty, honor, country.

In July 2015, they started their site visits. The Cadets gathered information, provided a phase one draft plan, and reconvened after the beginning of the new school year, as Firsties. They drove on, motivated, enthusiastic to accomplish the mission. From time to time, part of my job was to provide feedback to support the Engineering Department cadet course evaluations. I commented on their grasp and understanding of the mission and that their technical expertise in engineering and leadership, in a relatively short time, was incredible. Reports from the kennels were resounding. At each kennel these young cadets visited, they easily interacted, displayed confidence, explained the mission, ensured inclusion of the MWD Kennel Master and Handlers, and probably learned more about MWDs than they wanted. What they did was lead the way for a program where the end state was "improved kennel designs." Their presence and energy endeared them to every kennel they visited.

The draft design was completed in August and transferred into computer models. Their computer models provided a template and data that validated construction design, overall costs broken down, per kennel size: construction, materials, sustainment, utilities, waste removal. Designs outlined, structural loads, environmental impact (measured decibels), and state of the art materials with extended shelf life. I'm not an engineer by any means and while I understood their template, I would need to be briefed on the technical aspects in layman terms.

September was the moment we'd all been waiting for. The previous year had been extremely rewarding. We established an unknown resource and stakeholder in West Point. We now had a fully embedded partnership to assist the canine mission that could possibly change the facades of our global MWD platform. West Point was now an official stakeholder of the Army MWD program. The joint efforts of veterinarians, the Cadets and their advisors pulled together three prototype designs.

The matrix design computer model was a three-dimensional projection of each of the three kennels. Each design was intricately framed. The three-dimensional building designs gave us the ability to follow from the ground up each phase of construction from start to completion. With each slide we stopped, the cadets briefed on design methodology, which included sloped concrete for waste removal, improved LED lighting fixtures, and state of the art materials with an estimated shelf life of fifty years.

Even more astounding was the efforts by these cadets, ably assisted by their advisors. Each of the three kennels was designed to meet building codes and the costs of materials were broken down to the last nail. The design was environmentally sound, and one of the recommendations to support a more ergonomic flavor was to have the roof covered in sod, with the ability to grow plants. A fine idea, but we decided that solar panels would best fit the energy "green" ergonomic requirement. I couldn't see a handler pulling general maintenance mowing the "roof." The design would have provided natural year-round insulation saving hundreds of operating maintenance dollars.

Our vets came up with a recycling compactor that was environmentally a vision of beauty! As the kennels amass a sizable amount of excrement, the Vets designed a compactor where the excrement would be washed from the kennels into a compactor. The compactor would "recycle," and the recycled excrement somehow used as an external fuel to help power the solar panels. We ran a test with the Connecticut Army National Guard, and it worked extremely well. The cost to procure, obtain approval from the Army Maneuver Support Center of Excellence (MSCOE), the Army agency that approves new equipment, never received the support to bring it into the Army inventory. However, the design remains on the shelf and maybe as we strive to a "Green" political agenda may very well be resurrected.

This was the epitome of "Army Strong." One team, One fight. When these young Cadets went on to serve in the Army, they were just a little more aware of this unique Army mission. Our canine community and Vets are more aware of the construction and design capabilities to maintain MWD facilities. Our handlers better understand the costs of construction materials and the need to maintain building codes and assess ergonomic means to project cost savings for the longevity of a kennel. Moving forward and maintaining these relationships is the key as we strategically look at the Army MWD and DoD global footprint.

Cost savings were less than had been projected. Under military construction (MILCON) caps at the time, anything over $1M was considered MILCON. Any costs under $1M were under major military repairs. The difference, time, and lines of accounting. Major Military repairs could be funded through installation annual budgets. MILCON was a five-year process where the installation had to present justifications to the Army to compete for other Army wide MILCON projects from the very small pool of appropriations provided from the National Defense Authorization Act.

We had a singular objective to come up with a new static kennel design to replace our decades' old hand me downs. We came up with a state-of-the-art design that was durable and long lasting and considered environmental and geographic variables. Was it employed? No. The DoD did not endorse the new design, regardless of cost savings and enhanced longevity. The Army, at no fault of their own, attempted to use the new design for a small kennel at Fort Irwin California. Using the cost analysis, it came in just under a million dollars. But that was stopped in its tracks when the cost of labor drove construction costs up to almost $1.5 million.

The design now exists in someone's computer database. The Army made the effort that the DoD should have pursued as part of research and development. The way ahead always seems clear, but, unfortunately, as we have no champion in the DoD, we remain as our predecessors did during WWII on Bataan and Corregidor, the battling bastards of the canine program…No Mama, No Papa, No Uncle Sam. Will the DoD ever move forward to rebuild these antiquated kennels?

Chapter 32 – Gone but Never Forgotten

Staff Sergeant John Mariana and Bronco

In every war, we lose young men and women. None are acceptable losses. Each soul we lose has parents, wives, brothers, sisters, or significant others. They are all loved, and their lives are cherished. We can plan for combat, we can evaluate the numbers of troops, dogs, airplanes…all the tools of war necessary to accomplish the mission. But what we don't do is assess and pay particular attention to the war that goes on in the minds of these brave troops.

Troop leaders are mission focused, "Get the Job Done!" Or "Suck it up, Troop." Yes, leaders are trained to see the warning signs…and get troops help. But sometimes the warning signs are so entrenched or camouflaged, that leaders can't get them the help they need. The loss of two soldiers due to their emotional wars, two wonderful young men who I admired and loved, still haunt me today. As a leader, you always wonder, what more I could have done to take care of my kids? Here are their stories.

In October 2010, Sergeant John Mariana and Bronco were deployed to support Special Operations Forces (SOF) at Camp Vance in Afghanistan. Unlike our conventional MWD teams, there was no thirty-day acclimation. They were immediately sent into action with Special Forces' "Operational Detachment Alpha (ODA)" teams. These teams are the tip of the special forces—agile, mobile and self-contained. They operated nonstop during their deployment and were credited with finding over thirty explosive devices. They were canine detection team superstars.

I met John and Bronco during a visit to evaluate the canine program in June 2011. John and I quickly bonded. We were two Brooklyn boys and die-hard New York Yankees' fans! A perfect match. He was a hard charger—and I knew that team would take care of their ODA.

The Special Forces Kennel Master at the time was an MP, SSG Kyle Slania. He was responsible for the joint Special Forces Multi-Purpose Canines and traditional Military Police Specialized Search Dogs and Patrol Explosive Detection dogs. His primary mission when the Colonel walked into the kennel… put a diet coke in his hand. It was a tradition that the SOF kennel achieved with glory and distinction. I always felt privileged visiting the kennels. As with most units, the enlisted folks like to remain within their "family" and an Officer's presence is "tolerated" but unwanted most of the time. I was fortunate as I always felt accepted. Allowed to blend in with the handlers, if anyone had a question, or wanted to bitch, the floor was always open. This was a free fire zone. Anything said in the kennels, stayed in the kennels. I was not going to violate the privilege they'd extended to me.

As part of the Special Operations team, SSG Slania monitored their MEDEVAC frequency so he could track incoming wounded canines. Special Operations Groups had a Veterinary team that was co-located next to the Combat Support Hospital on Bagram AB. The day I visited, word came over the net that a medevac was bringing in a wounded MWD (Bronco) and handler (Mariana). Both had been wounded. I jumped into the vehicle with Kyle and another handler, and we sped off to the MEDEVAC landing site. We arrived as the Blackhawk was landing—and the vets and medical staff rushed out to get the dog and handler to surgery. Bronco's wounds were less severe than John's, receiving shrapnel wounds to his stomach and taken a round in his snout. John's abdominal wounds were more serious.

There is no stronger bond than the physical and emotional one Bronco had with John. As with all MWD teams, the two of them would lay down their lives for each other. After Bronco had undergone surgery and the anesthesia started to wear off, he became restless, whimpering for his "daddy." John had come through surgery well but had to have his spleen removed. Bronco was inconsolable. He just wouldn't stay still. As rambunctious as he was, the vets didn't want to sedate him.

So, we organized a covert operation. The hospital was connected to the vet clinic, and easily accessible. With a "liberated" gurney, the vets strapped Bronco on. Once they had him covered him up, off we went, venturing into the Combat Support Hospital on a mission of mercy. We got lots of suspicious stares from the medical staff as we maneuvered through the corridors but remained undetected. When we finally reached John's room, he was still groggy from surgery. However, once Bronco saw his beloved "daddy," he leaped into John's arms. Love is always the best medicine! We all stood there sobbing tears of joy.

Eventually, a surgical nurse appeared. She shook her head but allowed John a few more minutes to enjoy Bronco before we whisked him back to the Vet Clinic. John was going to need some time to recover and Bronco's status as an explosive detection dog would have to be evaluated. The medical disposition, Bronco was sent back to Lackland AFB for medical evaluation at the Holland Veterinary Hospital…and John returned to Ft. Carson, Colorado

Fortunately, based on the clinical evaluation it was determined that Bronco's wounds were debilitating, and Bronco would be retired. That meant John could automatically adopt Bronco.! Roby's Law, a federal statute, states that if a Military Working Dog—or their handler— sustains wounds in combat, that handler may adopt their dog. I was happy for John. We stayed in touch…emailing each other from time to time and chatting on the phone about those "Bronx Bombers." However, there was something amiss with John. He seemed lost. Bronco's adoption was dragging on and his inquiries to Lackland provided little or no positive information on the status of the adoption.

I reached out to John's commander and found out that John was depressed and agitated. He took to sleeping in the kennels to feel close to his beloved Bronco. It sounded as if John might be suffering from PTSD…and we agreed that Bronco was the therapy John needed to heal. Reaching out to the Commander of the DoD Veterinary Hospital, we quickly found that the adoption paperwork had been misplaced. Once that was quickly corrected, Bronco's adoption was expedited. Within thirty days, John and Bronco were reunited. It had been almost nine months since John submitted the adoption application.

While this was a joyous occasion, it quickly turned sullen. After two months, Bronco began experiencing difficulties functioning. John brought him to the Vet and after running tests, an inoperable cancerous tumor was found. Bronco would have to be put down. I think that tragic news pushed John to the limit. He'd hoped that he would have years with Bronco, but it turned out to be just a few short few months. I think that it broke his heart. All the anxiety and pain of the separation, realizing that he now had to put Bronco down, was too much for him. Regardless of how his wife and young daughter demonstrated their support and love, I don't believe John was able to get over the loss of Bronco.

On November 28, 2012, SSG John Mariana, father, husband, soldier took his own life from an apparent overdose. He was thirty-three years old. An American hero, he survived being wounded, but those physical wounds healed. However, his emotional struggle, those demons were too deep. My pizan from Brooklyn, I took the news hard. Three deployments, Bronze Star for valor, purple heart. He was the epitome of an American hero…and a casualty of war and life's cruel circumstances. He did his duty, saved lives and hopefully is at peace, playing with Bronco in Heaven.

Staff Sergeant Raphael A. Futrell

In the Army, you are supposed to be guided by a chain of command…and there is respect for those of higher rank. In some cases, there is the smart ass who must push the envelope, and in my case, it was "Tony." Our first meeting set the stage for what would become a "father–son" relationship, cut off too soon by "life."

In the spring of 2009, my wife, the other LTC Vargus, and I were on leave in Hawaii. Beautiful weather, picture perfect beaches, sheer paradise! The Aloha atmosphere is truly relaxed and serene. At least that was the intent.

Schofield Barracks is home to the 8th MP Brigade, "The Guardians of the Pacific." Their immense area of responsibility covers the entire Pacific Rim and the Korean Peninsula. Like all the other MP Brigades, the 8th was tasked with providing MP assets to Iraq and Afghanistan. They too were doing more with less. The 520th MWD Detachment was scheduled to deploy a Patrol Explosive Detection Team (PEDD) to Iraq in November. Since we were on "Island", I wanted to get up there to introduce myself and provide the team with a situational update.

It felt good to be out of uniform. I easily adapted to Hawaiian shirts, shorts, and flip flops. I got up to Schofield to make my way over to the kennels. Of course, adorned with a New York Yankees baseball cap. I wandered around the base, looking for the kennels. Eventually, I found them behind the MP station. I knew the Kennel Master. He'd served as one of the KMs under SFC Peek in Iraq. He was a great young man, meticulous and competent. We chatted for a few minutes. I gave him a quick overview of what was going on in Iraq and Afghanistan…and then, he sent someone to get SSG Futrell.

It was no secret who I was. SSG Futrell had been briefed that the CENTCOM MWD Program Manager, a LTC was coming for a visit. Normally, regardless of whether you are in uniform or not, if you recognize the rank of a person, you render the greeting of the day and a courtesy salute. The senior is not required to return a salute when not in uniform, but traditionally does. I was expecting that this would be the way the introduction would go.

This smug kid walked into the office, stuck out his hand and said, "Hi POPS." The kennel master called him to attention and started to dress him down, but I intervened and added my own "special flavor" to our introduction. Informing him of my rank and that should he address me as "Pops" again, I would introduce the tip of my size ten combat boot to the entrance of his anal cavity and proceed to have it protrude through his sphincter and come out of his "friggin mouth!" All that delivered in one breath in my bellicose Brooklyn accent.

I believe I gave a pretty good performance…one worthy of being nominated for a MP golden globe award. All the while, I was trying to control my laughter at this kid having the gall to call me Pops. It was pretty ballsy. He was just trying to be a smart ass. I liked him immediately. He was a fine young man with a previous deployment to Iraq. We chatted for a while. There was something special about him. He knew his job, was eager…but not over eager…capable, and I could see that he was a leader not a follower.

He impressed me. I told him to let me know when he arrived safely in theater and stay in touch. Shortly after this brief conversation, he returned to his duties. As my wife and I were getting ready to depart, the Kennel Master turned to me and said, "I'll make sure your son reaches out when he deploys." I thought great, now I have another kid who's going to tap me for money!

When Tony deployed to Iraq, he was assigned to the central military region—Multinational Division—Central (Baghdad). He pulled missions in a very IED "rich" area. At the same time, the canine transit hub in Kuwait was moving approximately one hundred dog teams a month in and out of theater. They were super busy. Ali As Salem Air base, the MWD "Hub, "was barely keeping up. Shortly after SSG Futrell deployed, one of the Army NCO's supporting the Hub rotated back to the states. His replacement would not be arriving in theater for almost a month. A month without a full contingent at the hub was a no go. I'd been in contact with Tony, and as I said he was a leader. I reached out and called in a favor, asking the Operation Iraqi Freedom Canine Program Manager if I could "borrow" SSG Futrell to fill the gap until the Army replacement arrived. The "deal" was made. However, his commander had one caveat. If there was significant activity in Baghdad, SSG Futrell must be returned immediately. "Roger Sir,", I said. The deal was done. With his commander's approval Tony was detached to the Hub for thirty days.

He arrived at the hub and immersed himself in the task at hand. Because he was a Staff Sergeant, he complimented the Air Force NCO Team Chief, adding NCO continuity. His presence allowed the 24/7 operations to have an NCO on each shift. He was a superstar.

During SSG Futrell's time at the hub, I was scheduled to perform a Law Enforcement Program visit in Iraq and Afghanistan, giving me the opportunity to see him as I passed through. When I walked into the Hub, he immediately got up from his chair and gave me a big bear hug, no formalities, just a big, "HI POPS." I just shook my head, happy to see "my son."

My flight wasn't for another day, so I was able to see him in action…making it happen. He was a three-ring circus unto himself—juggling dogs, troops, working flight schedules, and providing updates to theater program managers. He was the real deal. We talked about what was next in his career, and I told him that when my wife and I came to Hawaii, I wanted him to bring his daughter over to Bellows for a day at the beach. We talked about his future in the Army, career plans, finishing his degree. I had taken him under my wing—and he was embedded into my family like he really was one of my kids.

The teasing included talking down to him and embarrassing him in front of his peers. And the icing on the cake was calling him to attention and making sure that he responded with the appropriate "Yes Sir" to my fatherly insults.

"So, SSG Futrell, you really are a dumb ass."

"Yes Sir!"

"Your dogs are smarter than you…right?"

"Yes Sir."

And on and on, it was good to have the upper hand for a change. All in good fun, but as one of my "children," as they all were, there was nothing that I wouldn't do to take care of Tony or his family if he asked. When you lead, you lead from the front, set the example, admit your mistakes, but never forget that your sole responsibility—always taking care of your troops, no matter the time, commitment or effort. I loved them all.

While I was there, things heated up in Iraq. SSG Futrell's command called and requested his return. The deal had always been that when that call came, he would return to his unit. However, it was going to leave a gap…not only for manning the hub, but in the leadership that Tony brought to the team. When I was getting ready to head off to Afghanistan, I had a revelation. As I was getting ready to head to the flight line, I pulled Tony aside and told him point blank, "I'm giving you a direct order…if you ever feel like doing something stupid, you are to call me or the other LTC Vargus, any time, 24/7, no BS, no excuses. If you don't, I'll kick your ass!" That was the last time I saw him—and to this day, I still feel the emptiness and anger of my failure as an officer and surrogate father.

I completed my mission and returned to CENTCOM. I was glad that I got to see Tony. Shortly after my return, I got an email from him letting me know he was okay. In his email, he asked my help arranging his mid-tour leave to coincide with his girlfriend's. I couldn't interfere with his chain of command. I told him that he needed to bring up his request through his chain, since my hands were tied. That's the military way, plus his command had already supported my request for his temporary assignment to the Hub.

I didn't think anything of it. It was unfortunate, but mission first. His demeanor seemed fine, he referred to me as "Pops" and all seemed well. But just a few days later, I received the news no "father" wants to hear.

On March 26, 2009, I was at CENTCOM in my "cube." Sometime in the early afternoon the phone rang. It was SFC Jeremy Peek, the Iraqi MWD Program Manager. His voice was sullen and immediately I knew something was wrong. There was a pause and then he said, "SGT Futrell died."

Everything came to a stop. I was speechless, I couldn't fathom or believe that he was gone. My first thought was that he had been killed on a mission. I could understand that. SFC Peek explained that Tony's death was not combat related. It was suicide. I ended the conversation and sat there motionless. I still remember the feeling of loss and despair as if this happened yesterday.

When I went home, I was distraught. The more I thought of it, the more I relived my last conversation with him. I distinctly gave him an order. In the Army you don't disobey orders, you carry out your orders. I had expected that Tony would follow his. They were clear, concise, and direct with no room for misinterpretation.

What would have caused him *not* to follow his orders? They were not a joke. I was serious as a heart attack. They were no s**t orders. If he had followed them, he would be alive, and I could have helped him. God, I could have done whatever it took to get his leadership to come to his assistance. Why…why the fuck didn't you do what the fuck you were told to do?

Psychologically, it's been difficult for me to process and accept the fact that he's gone. I've always felt that it was my personal failure as a leader that he took his life, because I must have failed to communicate my orders properly. I've beaten myself up repeatedly.

In 2012, when I was transitioning from Active Duty, I was driving up from Tampa to my new civilian job at the Pentagon. I contacted his mother and asked if I could visit and pay my respects. She reluctantly said yes. When I arrived at her house, she was gracious and kind. She showed me Tony's room and some of his childhood items. I felt uneasy and told his mom how I felt responsible for Tony's passing. She reassured me that it wasn't, letting me know that he was at peace. She took me to his gravesite. It was a surreal moment. I was filled with emotion, still hurt, angry.

As I stood there for a few seconds, I reminisced on the limited time that Tony had blessed my life. I wish he was still here. Shortly after we returned to her house, and feeling slightly relieved, I paid my respects and continued my way, wondering if this would bring me closure.

I think of him often…Pops misses you every day!

Chapter 33 – Sacrifice and Remembrance

While the bureaucratic battles ensue, the reality of war is that young men, women and dogs get injured. Some pay the ultimate sacrifice. Everything I've shared about the shortfalls and gaps in the program…from outdated policy to lack of funding…was precipitated on one driving principle…taking care of our troops and our tremendous canines. I've been accused of bypassing the chain of command, making independent decisions, being obnoxious, being uncooperative, and aggressive. Having "Gone Native" and being "Rogue." That makes me a typical New Yorker!

The useless crap of never-ending staffing, power point ranger nonsense and ridiculous charts and PhD level squiggly lines, in comparison to the responsibility for caring for our troops is unimportant. The only ones that believe this window dressing matters are lifelong…mostly politically appointed bureaucrats…who have never left the five-sided building. Concerned with policies and legal opines, from their cushy offices, they have no clue the impact their nice and tidy policies have on the young men and women who shoulder weapons and defend our nation.

At CENTCOM, you were charged with the responsibility of making informed decisions, passing subject matter expert recommendations up the chain. The difference between CENTCOM and the 5-sided asylum was that, in most cases as the subject matter expert, your recommendations were presented in face-to-face briefings with General Officers. No multiple layers of bureaucratic review. Decisions were made on the spot—not waiting for six months of bureaucratic staffing. You were expected to use your judgement and execute.

For those of you who I had the honor to serve with, those of you who happened to see me frustrated with people making ridiculous decisions—regardless of rank or position, my New York would come out! No one places "my kids" in harm's way unnecessarily. And if you don't know what you're talking about, I opened my mouth…and like a Shakespearean sonnet, I'd tell you, "Go fuck yourself!" Stupidity is no excuse for getting someone killed.

I was always in awe of our young men and women. Where do we get such brave heroes? They step forward and shoulder the mantel of this nation's security, selfless service and self-sacrifice their mantra. How fortunate we are that these wonderful, dedicated cream of the crop kids chose selfless service over self-gratification. They come from different backgrounds, races, lifestyles. Our Armed Forces, the melting pot. They are the ones who sacrifice for our freedoms. Two and four legged.

During my tenure at USCENTCOM, we lost thirty-two dogs and handlers. Army, Air Force, Marine, Navy, and Army Veterinarian. They were lost from indirect fire, IEDs and disease. Each life cut short is a tragedy. Every time a report came in from the field that we lost a canine hero, it was heart wrenching. I made it a priority when I was traveling throughout the CENTCOM AOR to attend every remembrance ceremony honoring a fallen MWD team. I wish the public back home could have seen the sheer emotion of these ceremonies. They celebrated a life given in service not only to our nation, but having laid down their lives, so others could live. Protecting their brothers and sisters in arms.

In Afghanistan, all fallen military personnel received a guard of honor, including MWDs. Once their remains were processed in at the Bagram Mortuary Affairs Unit, the remains were prepared for "dignified transfer "to Dover AFB. An all-hands message announced the funeral procession from the Mortuary to the flight line. Soldiers, sailors, airman, marines, civilians, contractors, coalition troops all mustered on the tarmac to render honors. Hundreds came out to pay their respects. The honor guard came to attention and rendered the hand salute as the remains were slowly driven to the awaiting aircraft for the final journey home. It was a solemn ceremony…and many wept.

On one of my trips to Afghanistan, I had the honor to participate in what is an indelible memory. Mortuary affairs announced a dignified remains ceremony for a Military Working Dog. I lined up on the tarmac with over a thousand joint and coalition partners who came to pay their respects. I was filled with emotion. A fallen hero, going home with the respect and affection of service brothers and sisters in arms. We came to attention… and as the small coffin was escorted to the waiting aircraft, tears filled my eyes. I stood at rigid attention, sad, yet extremely proud. For this hero dog, while small in stature, he was a giant of a soldier, and had given his life to save others.

Chapter 34 – Awards and Decorations

MWDs are not authorized to receive medals for outstanding service or acts of bravery. Officially, the 2017 National Defense Authorization Act addressed awards and decorations for MWDs. The primary reason…administrative processing. When a two-legged trooper is recommended for an award, the chain of command generates a recommendation which is submitted to the level of command having approval authority. The recommendation is reviewed and once approved, the award is processed through the service military personnel system, orders issued and posted to the respective trooper's military personnel record.

There is no personnel management system for dogs. They do not have service records per se. The current MWD record is maintained in the Working Dog Management System database (WDMS). It tracks each dog's training and missions from the time it is accepted as a DoD asset to the time it is disposed of at the end of its life cycle. The kennel master, trainer and handler record each element of a dog's service life cycle. Canines…regardless of the battles between animal groups and semantics of being referred to as equipment…are still procured and managed as military equipment. Equipment is not recognized for acts of bravery.

The military would have to establish a cross-over personnel system and realign awards and decorations regulations for working dogs. However, with the Working Dog Management System it would have been a labor of love to add an awards and decorations section in each of the services WDMS portals. The recommendation process would be the same as "human awards." A recommendation for an MWD award would be forwarded through the chain of command, an awards board, and issuance of permanent orders maintained by the issuing headquarters and posted to WDMS. A simple process. We need to be able to recognize our canine heroes for their bravery, preserving their distinguished service as their legacy for all time.

The 2017 NDAA directed DoD and Military Services recognitions through individually created certificates of merit. Each service was required to provide the procedures and a standard certificate in their respective MWD policies. But that was not to deny our dogs from being recognized unofficially.

Throughout our Nation's wars, dogs have been awarded medals for valor and been given rank. As far back as WWI, U.S. dogs were given rank. The most recognized War Dog…Sergeant Stubby, who received the Purple Heart, France's Grand War Medal, and campaign medals for his service in WWI…is forever remembered in a place of honor in the Smithsonian Institute. Chips, the WWII German Shepherd, was awarded the Distinguished Service Cross, Silver Star and Purple heart for bravery. And there is Sergeant Reckless, the Marine Pack Horse, whose exploits of bravery earned her a Purple Heart and Presidential Unit Citation. Later, all were revoked by Congress.

Yet the traditions of presenting unofficial awards remains. Units have "created" canine purple hearts and kennels and handlers promoting their MWDs into the NCO ranks. These prestigious honors are presented with all the pomp and ceremony they deserve. The debate continues, with the hope that Congress and the Executive Agent will actively reverse this injustice and recognize their service and bravery.

PDSA Dickens Medal – Animals in War and Peace

In 1943, the British established the PDSA (Peoples Dispensary for Sick Animals) Dickens Medal. The PDSA Dickens medal was instituted to recognize outstanding acts of gallantry and devotion displayed by animals while serving with the British Armed Forces or Civil Defense units. Recognized as the animals'

"Victoria Cross," it is a fitting tribute to the gallant animals "who also served" and whose remarkable contributions helped save lives. Stories of the British heroes are captured in Robin Hutton's book War Animals, The Unsung Heroes of WWII.

These medals are testaments to the diligent, resolute, fearless and relentless qualities shown by animals in action. To date, the PDSA medal has been presented by members of the Royal Family and Senior Military leaders to thirty-two pigeons, thirty-four dogs, four horses and a cat.

Here in the United States that concept has been adopted to continue the PDSA tradition. The Animals in War and Peace Medal of Bravery and Meritorious service, recognized by Congressional Resolution honors our gallant Military and First Responder canines. In November 2019, the first "Animals in War in Peace – Medal of Bravery" awards were presented at the Rayburn Congressional Office Building. The event honored U.S. animal heroes past and present. Listed below are just some of the awardees from that ceremony:

- Marine Equine "SSGT Reckless" whose bravery during the Korean War saved hundreds of lives.
- Homing Pigeon "Cher Ami," the pigeon who although severely wounded returned to her perch with the message from the Lost Battalion in WWI to stop the friendly artillery barrage on their position.
- Marine MWD Scout Dog "Stormy" who saved countless lives in Vietnam alerting on enemy positions, tunnels and detecting punji pits and explosives bobby traps.
- Marine MWD Specialized Search Dog "Luca," deployed to detect explosives, Luca was credited with over forty finds. Luca was wounded in Afghanistan losing her right leg and honored as the 67th recipient of the PDSA Dicken Medal.
- FDNY Arson Dog "Bucca," at the time in 2017, was the only Arson dog in the FDNY. Bucca was named in honor of Fire Marshal Ronald Bucca, FDNY, who perished on 9/11 at the World Trade Center. Bucca has worked more than one hundred fifty fires, twenty-nine homicides and five hate crimes. And Bucca was directly responsible for the indication leading to a confession and arrest in a double fatal fire.

COVID pushed the 2020 awards off to March 2022. Military and first responder animals were again recognized for their bravery and meritorious service. Their heroics and service forever captured in the animal annals of history.

- "Nemo" — Air Force, Vietnam, German Shepherd —In 1966, while on sentry patrol, Nemo alerted his unit of a group of enemy commandoes hiding in the brush. As Nemo charged them, he was shot twice: once in the nose, and then under his right eye which exited his mouth. His handler was also shot twice, and even though Nemo was wounded, he crawled across his handler and protected him until help arrived.
- "Ziggy" — Marine Corps Multi-Purpose Canine, Afghanistan/Iraq, German Shepherd — Deployed five times to four separate countries, Ziggy conducted over fifty heliborne direct action raids and assisted in discovering forty-three weapons caches, twelve homemade explosive facilities, one vehicle-borne IED, 330 anti-personnel IEDs and over 5,000 pounds of homemade explosives.
- "Cairo"—U.S. Navy SEAL, Afghanistan, Belgian Malinois — In 2011, Cairo took part in "Operation Neptune Spear," the raid that eliminated Osama bin Laden in Pakistan. Cairo

provided perimeter security during the raid, searching for bombs and escape tunnels. During his tours, Cairo was shot twice in the line of duty, recovered, and continued serving.

- "Feco" — Homeland Security, U.S. Coast Guard, San Francisco, CA, Hungarian Viszla — Feco is part of the Canine Explosive Detection Team at Maritime Safety & Security Team San Francisco. Still serving, Feco has tirelessly conducted over 1335 searches, including vehicles, cargo vessels and ferries, buildings, luggage, people, and mass transit facilities. Feco spent over 2335 hours conducting more than 584 demanding patrols around the Bay area protecting the California coastline.
- "Hurricane" — U.S. Secret Service, Belgian Malinois— In 2014, Hurricane protected President Barrack Obama and the first family when he stopped an intruder who breached the White House perimeter fence racing towards the residence. Despite being kicked and punched, Hurricane held the intruder captive until armed officers apprehended him. In 2019, he received Great Britain's PDSA Order of Merit (Britain's medal for Outstanding Devotion and Service to Society).

Other organizations work diligently to recognize our canine heroes. The American Humane Society recognizes dogs annually in the categories of Law Enforcement and Detection, Military, Therapy, Service, Shelter, Search and Rescue. Congress officially recognizes March 10 as K9 Veterans Day.

The Animals in War and Peace Award continues to receive accolades throughout the canine community. Attended by members of Congress, the military services, civilian first responders, our military veterinarians it is a prestigious event. There was one prominent individual missing…the DoD Executive Agent.

Sons and Daughters

There are many people who have been instrumental in my K9 career, professionally and personally. I've been blessed to have worked with many dedicated leaders of the canine community. Men and women from civilian law enforcement, the military and coalition nations (NATO). They provided me with the inspiration to go forth, working to improve the Canine Program. In my many years of service several hard chargers remain a constant in my life. They became affectionately known as my sons and daughter. Like my own children, destined to haunt me. To them I'm affectionately known as Pops…with love and affection.

"Den and Duane"

Hardenio "Den" Abdon, Master Sergeant, U.S. Army (ret). There must have been something in the water that Army Staff Sergeants drink. In 2009, I traveled to Afghanistan to meet with our law enforcement leadership team. The standard practice for visiting "dignitaries" was that we would be housed in the VIP quarters. Well, as a traveling LTC, I was far down on the priority list for special dignitary quarters. However, I was eligible for a cot in the dignitary tent at the Airfield in Bagram.

That wasn't going to be conducive for a crusty old man. After my initial meeting with the Provost Marshal, their MWD Program Manager called the kennels, getting my protégé MSG Jim Inniger and me a room "at the inn." Standard accommodations for most of the troops were "Chue's," plywood structures divided into rooms. They are easy to construct and include the basic amenities—a bed and a few nails in the wall to hang your clothes. Any other creature comforts were provided on the whim of the owner. It certainly beat a cot in an open-air tent, with one hundred other "guests."

As we made our way down to the kennels, this young wise ass stepped out, extended his hand and introduced himself. After exchanging greetings, I inquired where our quarters were. He blurted out, "Follow me, Pops!" I stopped dead in my tracks. POPS? Did I look like I had birthed this young whippersnapper? I knew I hadn't changed his diapers or burped him over my shoulder. I must have not gotten the memo about him being my long-lost son.

I played along, and that day established a lifelong love for a young man that I thank God for every day. I responded, "SON, I think the next thing you're going to experience is Daddy's boot going up your ass!" However, I couldn't be serious, and we both laughed. With his indelible good humor and huge smile, I just couldn't be angry. From that moment on, we became known in the Army canine community as father and son. Whatever the assignment whenever he would call the office, whoever answered the phone would let me know that "your son is on the phone."

As an NCO, Den is an astute young man. He envisioned the operational, strategic and interoperability of dogs. He understood the socio-political issues and how they affect military operations. He used these skills to build relationships and attend conferences. And he was respected by the senior leaders in our international canine community. When together, we nonchalantly introduced ourselves to the other attendees as father and son and became internationally known. And everyone believed it.

Every assignment he ever had…as a course developer at the U.S Army MP School, or as a Kennel Master and U.S. Army Europe MWD Program Manager, he brought his energizer bunny motivation. Den saw the need to infuse MWD training into NCO and Officer development courses. He taught MP leaders the tactics, techniques, and procedures to utilize and integrate MWDs before they appear on your unit deployment doorstep. He learned their capabilities and their function to best employ them as unit force multipliers. I still reach out to Den frequently. He never accepts the status quo, he's always ten steps ahead. Realizing that to beat our adversaries we must think outside the box, every second, every minute, every hour of every day.

Duane Stinson, retired Air Force MWD handler, trainer, kennel master. He was Dens' alter ego. Where there was one, there was the other. Duane…even though he came from the Air Force…he became a force multiplier for the Army MWD Program. He and Den were clones. With their strategic vision and innovative spirit, they turned the U.S. Army European program around. Certifications, operational readiness, mentorship, and their participation on the NATO MWD Expert Panel solidified their legend in the program. They attained international recognition as a canine program of excellence. They didn't sit back. They made themselves known by coordinating training opportunities, especially through NATO collaborations. They attained positions for U.S. Army handlers to attend the prestigious NORDEFCO, the Norwegian Defense Forces "real world" regional training exercise.

As we built NATO collaborations, we realized that terrorists don't limit their attacks to the military. NATO military canine units are also dual hatted, executing civilian law enforcement missions. NATO reflected the expanding need for the inclusion of law enforcement stakeholders, extending U.S. membership to the FBI and ATF. We were no longer polarized.

Through the FBI…yes, through my mentor Dave Kontny…the DoD joined the European Union counter-IED team. From an operational perspective, my duties were in the Pentagon. But we had a team with the strategic vision in Europe—Den and Duane. We tasked them to take the lead on behalf of the Army to participate in the EU meetings and they set into motion new joint collaborations. To say that my "sons" made a tremendous impression would be a miscarriage of justice. Their efforts, representing not only the Army, but the entire DoD MWD program, identified and solved a crucial training gap for U.S. military dogs in Europe. Their efforts forged a partnership with the Belgian Federal Police to receive imprinting on Home Made Explosives. A new partnership with Angelo Araujo with the Portuguese Federal police introduced us to their protocols of training with 500 pounds of homemade explosives. Those partnerships continue today. They further provided EU—NATO teams with presentations on odor detection, kennel maintenance, basic first aid and set in motion European MWD competitions.

As the "dynamic duo" worked their magic, there was a resurgence of information sharing. Information on how dogs were utilized during terrorist bombings in Belgium, France, and Germany. Opening the door to receive real world canine intelligence. Like so many others, they were Pioneers in the Army MWD Program.

"Nick"

SFC Nick Briggs (ret) was a ball of fire. He was the kennel master at Fort Jackson when I first met him. Knowledgeable, led from the front and held his kennel to standard. Nick was handpicked as the first MWD NCO for the National Security Special Event (NSSE) mentee program. The Army was taking the initiative to implement a train the trainer program. Selecting primarily, Kennel Masters, to learn the duties and responsibilities of the NSSE MWD Liaison NCO. Why was this so important. There was no formal training to perform this high visibility mission. We needed to start training a cadre of soldiers prepared to perform the mission. Nick was the unanimous selection. His tenacity and can do attitude, ability to assess the missions and execute, earned him accolades from the Military Liaison and Secret Service. There was no doubt Nick was a shining star. He rolled right into the job, and besides being introduced to the cast of USSS canine and the Army MWD Liaison, he jumped in with two feet. He took charge, coordinating missions, maintaining accountability of the more than one hundred MWD teams. He reset the negative perception at the time held by the USSS. His after-action report was spot on, he identified the lack of planning time, with recommendations for future mentees.

Nick also drove the train, hosting the test of the FBI coordinated effort to train MWD's on TATP and HMTD. Along with our great FBI Special Agent Bomb Tech, he wrote the script for our DoD – FBI partnership. As my other son, he was closer that Den and Duane, and if I needed to get an opinion in a hurry, I could always count on Nick to give me an operational and strategic answer. We spent some time in Hawaii, before he left the Army, as always, he was doing great things, creating partnerships and training with law enforcement and coalition partners.

"V"

I first met Sergeant Major (SGM) Viridiana LaValle at Fort Stewart, Georgia, in August 2012. I was in the process of retiring, "the second time," and accepted an invitation to visit the kennels for the sole purpose of visiting superstar Special Operations Kennel Master Kyle Slania. Recently returned from Afghanistan. I was pleasantly surprised when I was also introduced to SGM LaValle. From the second I met her I realized she was a ball of energy. At the time, SGM LaValle was the Fort Stewart Kennel Master. A soldier's soldier, in fantastic physical shape, confident, motivated, technically proficient, doing what leaders do…setting the example and leading from the front.

From the onset, I saw that she took no crap. She set the standard by her personal appearance and competency as Kennel Master…and she expected the same one hundred percent commitment from her handlers. She never wavered. I was impressed and have remained impressed since that day.

She was the bearer of good news, or maybe not, informing me of the position vacancy at the Army Office of Provost Marshal General for the MWD Program Manager. If it wasn't for her, my career in the MWD program would have probably ended. She did for me what she does for her soldiers. She was looking after one of her troops. Our friendship, both professionally and personally, has been a bond of understanding, reflection, envisioning short- and long-range goals for the Army MWD program. Our priorities. Like Den and Duane, her vision has taken the Army MWD Program to new levels of operational and strategic excellence.

When you're tough as nails, but fair, expecting that standards be met can sometimes lead to consternation and resentment within the ranks. The MWD program readiness is based on the numbers of certified MWD teams. The Army MWD certification is the toughest of the services, yet the annual rates vary for multiple reasons. Numbers of dogs and handlers on hand, deployed teams, time on station, personnel rotations, medical condition—all are underlying causes effecting certification. Anyone of these variables could tip the balance on certification percentages.

One of the issues was the personnel assignment system. Before the introduction of the 31K Canine handler Military Occupational Specialty (MOS), handlers came from the ranks of the MP units, and were normally an E4/E5. As deployments were extremely robust during OIF and OEF, MWD teams were on a merry-go-round. Gone a year, back a year. For a seasoned handler there was no path for promotion in the MWD program. Promotion to the E7 and above grades meant their only option was to leave the MWD program and return to the MP Corps, leaving a gap in canine handlers and exacerbating certifications.

The personnel system created a unique rollercoaster of hills and valleys, especially for our kennels in Korea. Korean tours were a year assignment. Dogs were installation property and normally remained in Korea for their service life cycle. Handler management was another detractor. MWD handlers were assigned to U.S. Army Korea for a year.

As handlers rotate into Korea, they are required to go through a forty-five day in-processing program…and then, at approximately the nine-month mark, troops begin preparing for their return to the states, culminating with another forty-five-day out-processing. Actual mission availability was six to eight months. In this slim margin of time on dog, they had to bond with a new dog, train up and certify. By the time a handler and dog were achieving team proficiencies, the handler was starting their out-processing. With these detractors, kennel masters were performing juggling acts to get MWD teams certified. The six to eight months estimate of time on MWD was dependent on multiple variables.

Statistical analysis of Korean certifications demonstrated that the best certification rates were achieved mid tour, when MWD teams reached the pinnacle of continuity. But training time to maintain their certifications needed to be balanced between MWD time and installation taskings. Prior to SGM LaValle arriving on the "Penn," as expected certification rates followed the standard "Penn" hills and valleys. But when LaValle assumed the U.S. Army Korea Program Manager position, certification rates ebbed throughout her year assignment. These less than acceptable certification rates were questioned by those in the Army MWD Chain of Command. Failure to train or holding handlers to standard?

It's tough to stand your ground. And it's tough to stay true to yourself. SGM LaValle, throughout her entire time within the MWD program, never compromised. She never wavered or lowered her personal or professional standards. What you see is what you get. She never accepted second best. Yes, certifications were low in Korea. SGM LaValle demanded nothing less than compliance to certification standards. She never shirked her responsibilities, and was right down there in the trenches mentoring, working with handlers, helping them achieve their professional MWD team certification. There are no do-overs if a canine misses an IED.

I observed her leading from the front during my visit to Fort Stewart in 2012, demonstrating patrol techniques, donning the red man suit (bite suit) and demonstrating the patrol effectiveness of an MWD. To enhance certification, she recommended that personnel management of 31K handlers to Korea be changed from a personnel assignment…to a team deployment. Eliminating the certification problem. MWD deployments mandated that MWD teams be certified prior to deployment. Versus handlers on assignment, hopefully completing the arduous train up process and certify within the limited time they had in the kennels. The recommendation would have driven mundane certification rates of 30 to 50% to one hundred percent almost immediately. But U.S. Forces Korea leadership believed that changing the process would result in the elimination of the MWD Detachment (kennels) from their force structure. Accepting low certification rates over operational readiness. Her recommendations to add value to the Army program in Korea, but changing from an assignment to a deployment cycle would have revolutionized the program. MWD teams would have arrived certified, eliminating the administrative processing time.

She didn't shy away from making hard decisions. During her tour in Afghanistan as the Theater MWD Program Manager, she had the oversight of the non-traditional TEDD program. She wasn't swayed by Commanders demanding their TEDD teams. She put the TEDD's through their acclimation training. If they passed, she would provide her seal of approval. If they didn't meet the theater validation, they were not being released to their unit. They were sent back to North Carolina. She stood her ground and took heat from senior leaders in the chain of command. It was simple. If you were trained, meeting standards and demonstrated proficiency there was no issue. If you were not proficient, LaValle would not let you place soldiers' lives in jeopardy, period.

She also became the Afghan TEDD guardian angel. When the TEDD program was unexpectedly terminated, we had 42 TEDD's in Afghanistan, with no home to return to. Working to make sure no TEDD was left behind, she masterminded the process, working tirelessly to coordinate the adoption paperwork, and getting TEDD's their adoption physicals. Just another day, but her engagement, made sure we didn't have a disaster when the dogs returned to North Carolina. If the adoptions hadn't been executed in Afghanistan, these TEDD's would have returned to the U.S and would have been homeless!

Every TEDD that returned to the U.S. was adopted to their handler or transferred to the U.S. Capitol Police. This was an astronomical job. SGM LaValle, she was and is, "Of the Troops, For the Troops."

Her planning and coordination with the veterinarians and the Secret Service left no TEDD behind. When the TEDD program terminated, we still had forty dogs in Afghanistan. As no TEDDs were being returned to K2, all the documentation, physicals had to be executed in Afghanistan. She executed the entire process. Veterinarians in Afghanistan would perform the necessary health assessments and execute adoption papers. This was with the caveat that once the team arrived back on U.S. soil, the handler would sign the "hold harmless" agreement and the TEDD handler became the proud parent of an Army TEDD. Every soldier who wanted to adopt a TEDD was able to, because of the instrumental flawless adoption process facilitated by this great NCO. She masterminded a process in days, effective and efficient. To her, it was just another day at the office.

Attention to detail, holding herself and her subordinates to high standards, leading from the front and setting the example, earned SGM LaValle the distinction of being the first female soldier in the 31K career field to be promoted to the rank of Sergeant Major. The right selection for the right "Soldier." A demonstrated stand out leader. A leader whom I admire and am honored to have worked with over the years, a soldier who I'd follow without hesitation. Her resilience and mentorship, both professionally and personally, have been an inspiration for Army, NATO, and the Joint Service Military Working Dog community in War and Peace. Like my MWD sons, my heart bursts with pride for mi Hija (daughter).

Chapter 35 – Reflection - Equality

Dogs don't judge handlers by their ethnicity, race, religion or how they attained U.S. citizenship. They bond and love their handler, male or female, transgender or neutral. In today's divided climate, the new normal seems to judge a person's place in life based on their skin color, not their character or capabilities. But that doesn't faze a dog. Their love is unconditional and given openly to the person who earns their trust. Once developed, that bond lasts for a lifetime.

Leave your baggage at the kennel gate or you won't make it in the MWD program. Devotion to your mission and your four-legged slobbering fur ball is the measure of success. Anyone meeting the basic criteria…and is command selected…might become a member of the canine team. It's a very small fraternity. You *earn* the right and privilege to be a dog handler! And the work is unending. It's a commitment.

You won't make it if you don't have the ability. There are no perceived quotas, there can't be. This is an even playing field. A substandard handler can lead to injury or death for those you've been tasked to protect. Mission failure is not an option. You'll find that you are now included in a unique team of handlers. And every handler—military, U.S., foreign, police, first responder, FEMA S&R— is part of your family and they will have your back.

As you're trained to read and understand the behavior of your dog, so is your dog absorbing and reading the changes in your behavior. Whatever happens in your personal life needs to be checked at the door, lives depend on it. Psychology is an important part of your MWD service.

When you report to your command, or are deployed to support a USSS Protection mission, you perform your mission. The leaders of this nation are dependent on you…that young man or woman…tethered to your canine hero. The men and women who you are supporting don't care or are prejudiced about your ethnicity or race they are thankful and appreciative that a canine team is going to keep them alive. Your mission is not influenced on the stage of political rhetoric, nor politically directed media selling you opinionated nonsense. You and your canine are the only thing that separates young men and women from life and death. Dramatic, maybe, but truthful.

Using a beaten-to-death cliché, we are a "Band of Brothers/Sisters." We are reliant on each other, we live together, we fight together, we share our pain and sorrow together. We are the only ones we have. No one else understands the bond, no one else understands that the love we share for each other crosses all boundaries.

It's a tough club to get into, but the rewards are worth the hard work, dedication to the mission, your canine partner, unit, and nation. You're taxed physically and mentally, as you become both caretaker, mom and dad and animal psychologist. Every day is an Olympic event. It's your drive and desire that you bring to the canine team that makes the difference, not the colors of the rainbow. No one cares. I've stood shoulder to shoulder with my canine family and hope and pray that I was able to support the MWD and first responder community.

I have closed out my government service, proud to have been labeled Rogue and having gone native…it's an honor that tells me that I didn't accept complacency. My challenge to you, do the same, lives depend on it. Drive On!!

In Canine Confiduris

I wish I could write a story about the men and women who impacted my "canine career" and give them the just reward and recognition for their mentorship and friendship. The spectacular troops they were and are. They made tough decisions to take care of our troops and their four-legged heroes. I know I didn't do justice to recognize everyone who impacted my life. You made the difference.

Tributes

Like General of the Army Douglas MacArthur, I too am the old soldier who never dies. I'm the thorn in your side that won't go away. In my role as a leader, I've been guided by many "American Idols."

First Sergeant Nicholas Aleshin, A Battery 244th Air Defense Artillery was not only a First Sergeant, but a mentor and surrogate dad. I remember when he ORDERED me to go to Officer Candidate School. Tough and relenting. But mostly I remember the day I graduated OCS and the tears he had in his eyes, the pride he had, and how honored I was to receive my first salute from him. I loved that man and credit him providing me with the foundations of my military career.

Sergeant Major Nicholas Braccia, my other "Dad". I was an upcoming young Captain and Nick was the Group Operations Sergeant Major. As a young hard charger, newly minted Captain in a Group Operations section, I was just a little reckless. Nick saw the need for "counseling." Nick and I were from Brooklyn—we both had an indomitable New York accent. Many times, I would be summoned to my office for a "counseling" session with my "Dad." Some chats, some more corrective. Always with respect and always to help improve my ability to lead.

Both Nicks were clear about the only job of a leader… "Take care of your troops." You always looked after your troops before yourself. If you played the rank game as far as they were concerned, you weren't fit to wear the uniform.

The Sergeant Major was my shepherd, not only in the military but throughout my life. I always sought his counsel, and he was always there to encourage, guide and mentor. Years later, after I had returned from Iraq and was assigned to CENTCOM, my wife and I visited Nick in "The Villages," a massive retirement community in central Florida. By then he was in failing health, his overbearing presence dwindling. I hated to see this man fading away. In one of our last visits, he mustered the strength, calling me to his bedside, with tears in his eyes, and his ever-present Brooklyn accent looked at me and simply said, "Richie I'm very proud of you." That meant more to me than any award or decoration that I could have received. It meant more to know that in his eyes, I was a good son and good soldier. That was the last time I saw Nick, but I still carry his memory with me everywhere I go.

I've tried to mention the people, military and civilian, who molded my life, who have been the stewards of the MWD program. I may one day write a book on these great troops, but all I can do now is try to say thank you to those troops that tolerated me, mentored me, and never lost faith in the vision to improve the canine program during the past twenty-five years. I love all of you.

To those who I served with in the trying days at CENTCOM, the Army MWD program, Federal and Civilian Law Enforcement, the NYPD and FDNY, Animals in War and Peace, all the handlers who saved countless lives, from the bottom of my heart…Thank you. You are all family, and heroes in my eyes.

The final tribute…to my family. My wife, and kids. I can't even sum up in words how much your love and support has meant to me. You supported my decision to return to active duty, you encouraged me to do what I needed to do. We switched roles somewhere over the years, where you became my advisors and mentors. Thank you for your love and inspiration.

These bonds we share are not understood by those who have never worn the cloth of our nation. To ask a person to place their life in another's hands, purely because that is what your nation asks of you, defies explanation. Yet young men and women answer the call, they come from all backgrounds, all nationalities, colors, races, ethnicities, and together we serve. The saying goes that there are no atheists in fox holes, that's true. Bravery and courage know no color boundaries. When you're in a fight, you realize one thing…that everyone who serves this nation will lay their lives down to save yours. Until the day you leave this earth, your loyalty and devotion to your brothers and sisters will be steadfast.

God Bless the USA!

Glossary

9/11	September 11, 2001
AFB	Air Force Base
AFMES	Armed Forces Medical Examination Service
AFSF	Air Force Security Forces
AKC	American Kennel Club
AOR	Area of Operation
ATF	Alcohol Tobacco and Firearms
AUS	Australia
BAA	Buy American Act
BCT	Brigade Combat Team
BLUF	Bottom Line Up Front
CAOC	Combined Air Operations Center
CBRN	Chemical Biological Radiological Nuclear
CDC	Center for Disease Control
CENTCOM	Central Command
CONUS	Continental United States
COTS	Commercial Off the Shelf (local procurement)
CTTSO	Counter Terrorism Technical Support Office
DATR	Defense Animal Training Regiment (UK)
DHA	Defense Health Agency
DMZ	Demilitarized Zone
DoD	Department of Defense
EA	Executive Agent (Air Force)
EDC	Explosive Detection Canine
EDD	Explosive Detection Dog
FDNY	Fire Department City of New York
Five Eyes	Pacific Coalition partners (Canada, United Kingdom, United States, Australia, New Zealand)
FSR	Field Service Representatives
HME	Home Made Explosives
HMTD	Hexamethylene Triperoxide Diamine
IEDD	Improvised Explosive Detector Dog (Marine)
IEDs	Improvised Explosive Devices
ITRO	Inter Service Training Organization
IWDBA	International Working Dog Breeding Association

JCIDS	Joint Capabilities Integrated and Development System
JROC	Joint Requirements Oversight Council
JSMWDC	Joint Services Military War Dog Committee (see Chapter 2)
JSSA	Joint Service Support Agreement
K9	"canine"
KM	Kennel Master
LNO	Liaison Officer
MACOM	Major Army Command
MAJCOM	Major Air Force Command
MEF	Marine Expeditionary Force
MOA	Memorandum of Agreement
MOPP	Mission Oriented Protective Posture
MWD	Military Working Dog
NATO	North Atlantic Treaty Organization
NCO	Non-Commissioned Officer
NDAA	National Defense Authorization Act
NDS	National Defense Strategy
NETF	National Explosive Task Force
NIH	National Institute of Health
NMS	National Military Strategy
NSSE	National Security Special Event
NYPD	New York City Police Department
NZ	New Zealand
OEF	Operation Enduring Freedom
OIF	Operation Iraqi Freedom
OPMG	Office of Provost Marshal General
PDSS	Pre-Deployment Site Survey
POTUS	President of the United States
PPBE	Planning Programming Budget Execution
PPE	Personal Protective Equipment
R&D	Research and Development
RFF	Request for Forces
SECDEF	Secretary of Defense
SECSTATE	Secretary of State
SOF	Special Operations Force
SSD	Specialized Search Dog
STRT	Specialty Requirements Requirement Team (Air Force)

TADD	Training Aid Deliver Device System
TATP	Triacetone Triperoxide
TDR	Trained Dog Requirement
TEDD	Tactical Explosive Detector Dogs (Army)
TMT	Tasker Management Tool
TRANSCOM	United States Transportation Command
TSA	Transportation Security Administration
TTPs	Tactics Techniques and Procedures
UK	United Kingdom
USSS	United States Secret Service
VPOTUS	Vice President of the United States
WDMS	Working Dog Management System
WWI	World War I
WWII	World War II
YPG	Yuma Proving Grounds

References

1. ABC News interview, 2011
2. *War, Police and Watch Dogs*, Edwin Richardson, 1910
3. *War Dogs Caines in Combat*, Michael G. Lemish, 1997
4. *War Animals*, Robin Hutton, 2019
5. *War Dogs*, Rebecca Frankel, 2014
6. *Dogs of War*, Shelia Keenan, 2013
7. *Navy Seal Dogs – My Tale of Training Dogs for Combat*, Mike Ritland, 2013
8. *Hero Dog Story*, National Geographic, June 2014
9. DoD Directive 5200.31E, DoD Military Working Dog (MWD) Program, March 29, 2006
10. Air Force Instruction (AFI) 31-126, Military Working Dog Program, 17 January 2019
11. Title 10, United States Code
12. DoD Directive 3025.12
13. Title 32, United States Code
14. Joint Publication 3-15.1. Counter Improvised Explosive Device Operations, January 2012

About the Author

Dr. (Col) Vargus initially retired in 1995 after twenty-four years of service in the United States Marine Corps and United States Army. Following retirement, he began an extended career in law enforcement. In 2005 he was recalled to active duty, deploying to Iraq and Afghanistan, serving at U.S. Central Command as the Chief of Law Enforcement. In 2012 Colonel Vargus was released from active duty. He entered federal civilian service as the Department of the Army Canine Program Manager, and in 2018 assumed the position of Department of Defense Canine Program Manager, a position he held until his retirement in April 2020.

Dr. Vargus is a recognized global leader in law enforcement.

He is a graduate of the Army Command and General Staff, Air War College, George Washington University National Senior Leadership Course and National Security Management Course. He holds a master's degree in National Defense Studies and a Doctorate in Public Administration.

www.ingramcontent.com/pod-product-compliance
Lightning Source LLC
Chambersburg PA
CBHW080411230426
43662CB00016B/2375